ATLAST

Computer Exercises
for Linear Algebra

ATLAST
Computer Exercises
for Linear Algebra

Steven Leon

Eugene Herman

Richard Faulkenberry

PRENTICE HALL Upper Saddle River, NJ 07458

Acquisition Editor: *George Lobell*
Supplement Acquisition Editor: *Audra Walsh*
Production Editor: *James Buckley*
Production Coordinator: *Alan Fischer*
Production Manager: *Joan Eurell*

©1996 by Prentice-Hall, Inc.
Simon & Schuster / A Viacom Company
Upper Saddle River, New Jersey 07458

Printed in the United States of America

10 9 8 7 6 5

ISBN 0-13-270273-8

Prentice-Hall International (UK) Limited, *London*
Prentice-Hall of Australia Pty. Limited, *Sydney*
Prentice-Hall Canada Inc., *Toronto*
Prentice-Hall Hispanoamericana, S.A., *Mexico*
Prentice-Hall of India Private Limited, *New Delhi*
Prentice-Hall of Japan, Inc., *Tokyo*
Simon & Schuster Asia Pte. Ltd., *Singapore*
Editora Prentice-Hall do Brasil, Ltda., *Rio de Janeiro*

To the memory of

Bill Larry Neal and Kermit Sigmon

Preface

This collection represents the best creative efforts of the more than 350 faculty members who participated in the thirteen ATLAST workshops offered between 1992 and 1995. Workshop participants designed computer exercises and projects suitable for use in undergraduate linear algebra courses. From the entire ATLAST database of materials, the editors have selected a comprehensive set of exercises covering all aspects of the first course in linear algebra. Each chapter is divided into two sections. The first section consists of shorter exercises and the second section consists of longer projects.

Unlike other manuals of computer exercises, this collection is a massive collaboration representing a wide variety of views. With so many submissions it is not surprising that there was a great deal of duplication. In such cases the versions appearing in this book are generally composites of the various versions that were submitted. Further editing was needed in order to maintain a consistent format and style for the book. Also, the editors encountered many strikingly original shorter exercises that provided keen insights which would benefit students even more if they could be explored in greater depth. Consequently, some shorter exercises have been expanded into longer projects.

Using Software to Teach Linear Algebra

The computer has revolutionized the way we teach mathematics. This is particularly true in the case of linear algebra. Modern software packages enable students to more easily explore the many applications of linear algebra and to discover through experimentation significant theoretical results. Perhaps most important of all, the computer provides students with a means of visualizing major course concepts such as linear systems, linear transformations, eigenvalues and eigenvectors, and orthogonality.

In recent years, the graphics capabilities of software packages have improved dramatically. It is now possible to incorporate into classroom presentations sophisticated visual models illustrating all of the major theoretical concepts of linear algebra. This allows students to gain geometrical intuition into the subject matter.

In the past, before personal and laptop computers were available, many interesting applied problems were too computationally difficult to solve in the classroom. Now, with the aid of software, it is possible to present a wide range of applications while spending essentially no time on computations. As a consequence, students are more readily able to appreciate what a powerful tool linear algebra really is and how central it is to solving virtually all applied mathematics problems.

The computer is also a powerful tool for discovering mathematics. With the aid of software packages like MATLAB®, one can quickly generate a wide variety of examples. This makes it easy (particularly if the examples are carefully chosen) for students to discover interesting and important results. Theorems become much more meaningful to the students if they are discovered in this manner. Once a result is discovered from empirical observations, students are then challenged either to prove it or to explain why it is true.

The MATLAB® Software Package

The primary goal of the ATLAST Project is to encourage and facilitate the use of software in teaching linear algebra. While there are many excellent mathematical software packages on the market, we believe that MATLAB is by far the best choice for use in teaching linear algebra. Matrices are the primary data object in MATLAB. Indeed, MATLAB is the one package that is almost completely built around matrices. Additionally, MATLAB has many routines for generating random matrices and various types of structured matrices. This makes it particularly easy to quickly generate interesting examples in the classroom. Although the exercises in the ATLAST database make use of a variety of software packages, all chosen for inclusion in this book have been adapted for use with MATLAB. A collection of MATLAB routines (M-files) has also been developed to accompany this book.

The ATLAST M-files

Many of the M-files that accompany this book are designed to give visual illustrations of important linear algebra concepts such as coordinate systems, linear transformations, and eigenvalues. Other M-files illustrate visual applications such as using linear transformations for computer animations or using matrix factorizations for digital imaging. Still other M-files can be used to generate special structured matrices. Students are then challenged to discover properties of the special matrices.

The entire collection of ATLAST M-files can be obtained either from the ATLAST Web page:

http://www.umassd.edu/SpecialPrograms/ATLAST/

or by anonymous ftp from

ftp.cs.gasou.edu/home/atlast/mfiles

These files are required for many of the exercises and projects in this book. The files are being updated periodically; so check the web page or ftp site regularly for the latest versions.

ATLAST Lesson Plans

The ATLAST workshops held during the summers of 1995 and 1996 focused on developing lesson plans for classroom presentations using ATLAST materials and software. These lesson plans incorporate the ingredients of discovery, applications, and visualization that are prevalent in the ATLAST exercises and projects. We plan to have versions of many of these lesson plans available on the ATLAST web page in the near future.

The History of the ATLAST Project

ATLAST is an acronym for Augmenting the Teaching of Linear Algebra through the use of Software Tools. The project originated as an outgrowth of the activities of the Education Committee of the International Linear Algebra Society (ILAS). Steven Leon, a member of that committee, received an ILAS seed grant to write a proposal for a project to promote the use of software in the teaching of linear algebra. At about the same time, an independent Linear Algebra Curriculum Study Group (LACSG) was formed under the support of the National Science Foundation. The LACSG came out with a number of recommendations for reforming linear algebra education. In particular, they recommended that the first course in linear algebra be taught from a matrix-oriented point of view. They also recommended the use of software in teaching of linear algebra. The original ATLAST proposal can thus be viewed as an immediate effort to implement the LACSG recommendations.

The ATLAST Project was funded by an Undergraduate Faculty Enhancement (UFE) grant from the Division of Undergraduate Education of the National Science Foundation. The original grant supported five workshops during the summer of 1992 and an additional five workshops the following summer. Each workshop was limited to thirty participants. During the first year there were 250 applicants for the 150 workshop slots available. Although it was necessary to turn away many highly qualified candidates at that time, the project was able to accommodate all of those who reapplied the following summer.

The ATLAST Project received supplemental funding from the National Science Foundation to run an advanced Developers Workshop at the University of California, San Diego, during the summer of 1994. Twenty-five former ATLAST participants were invited for an intensive four-day workshop to develop additional exercises for the ATLAST book. Participants worked in small groups to develop exercises on particular linear algebra topics. The projects were tested during the workshop and revised and improved during the month following the workshop. The San Diego workshop produced a remarkable set of exercises, a large percentage of which have been included in this book.

A second UFE grant allowed the ATLAST Project to continue through the 1996–97 school year. This grant supported five workshops during the summers of 1995 and 1996. Because of timing, we were only able to include a few of the exercises from the two 1995 workshops in this book and no exercises from the three 1996 workshops. One of the 1996 workshops was an invited Developers Workshop. The emphasis of the 1996 workshops was on the development of classroom lesson plans that make use ATLAST M-files to help students discover and visualize important linear algebra concepts. We hope to make much of the material developed in these later workshops available over the Internet and in future editions of this book.

The original ATLAST proposal expressed hope that the project would play a significant role in the linear algebra reform movement. In addition to developing computer exercises, all of the ATLAST workshops devoted time to discussing linear algebra curriculum and other matters related to the teaching of linear algebra. The workshops offered a perfect opportunity for linear algebra instructors to get together to share their ideas and experiences. Many of the participants continue to play an active role in reforming linear algebra education. Indeed, there have been at least seventy-five presentations by ATLAST participants (including presenters and coordinators) at regional and national mathematics meetings during the past five years.

Future Plans

NSF funding for the ATLAST project will continue through the 1996–97 academic year. During this period we plan to make ATLAST lesson plans available on our web page. These lesson plans will also be included in the next edition of the ATLAST book. Future editions of the ATLAST book will appear periodically in order to keep up with new features in future versions of MATLAB.

All royalties from the ATLAST book will be contributed to the UMASS Dartmouth Foundation. The Foundation has established an ATLAST account which will be used to support future activities related to linear algebra education.

Acknowledgements

The ATLAST Project has been supported by National Science Foundation grants DUE 9154149 and DUE 9455074 as part of their Undergraduate Faculty Enhancement Program. Any opinions expressed in this book are those of the editors and not necessarily those of the NSF. Special thanks are due to Elizabeth Teles and William Haver of the NSF for all the help and guidance they have provided throughout the duration of the ATLAST Project.

We also thank the International Linear Algebra Society for its support of the project. In particular, the editors would like to acknowledge David Carlson, San Diego State University, and Frank Uhlig, Auburn University, of the ILAS Education Committee for their help and encouragement in the planning of the ATLAST Project. Former ILAS President Hans Schneider,

University of Wisconsin, also deserves thanks for his support and encouragement.

The Mathworks, Inc. has assisted the ATLAST project from the very beginning by providing both software and technical support for the workshops. Special thanks are due to Cleve Moler for his advice and cooperation throughout and to Mike Fisher for his assistance.

The ATLAST project was a resounding success thanks primarily to the very professional job done by the workshop leaders. The following is the list of these individuals together with their university affiliation and the years they served as workshop leaders. Although we have listed only one leader per workshop, it should be noted that all of the individuals listed played an active role in the planning and presentation of the 1994 Developers Workshop at the University of California, San Diego.

ATLAST Workshop Leaders

Jane Day, San Jose State University (1992, 1993, 1995)
Eugene Herman, Grinnell College (1992, 1993)
David Hill, Temple University (1992, 1993, 1994, 1995, 1996)
Kermit Sigmon, University of Florida (1992, 1993)
Steven Leon, University of Massachusetts Dartmouth (1992, 1993, 1996)
Lila Roberts, Georgia Southern University (1996)

Each ATLAST workshop had a local coordinator. These individuals put a great deal of effort into coordinating local arrangements and to seeing that the workshops ran smoothly and on schedule. Nearly all of these coordinators attended and played active roles in the workshops. The fourteen host institutions for the ATLAST workshops were partners in the ATLAST project. They provided the necessary classroom and computer facilities and any technical assistance needed for the computer labs. The following is a list of the host institutions together with local coordinators.

University of Maryland – David Lay
Auburn University – Frank Uhlig
West Valley College – Joseph Kenstowicz
University of Wisconsin – Rod Smart and Larry Farnsworth
University of Wyoming – Benito Chen
Los Angeles Pierce College – Thomas McCutcheon
Georgia State University – Valerie Miller
College of William and Mary – Chi-Kwong Li

Michigan State University – Richard O. Hill
University of Houston-Downtown – Elias Deeba
University of California, San Diego – Al Shenk and James Bunch
Seattle University – Janet Mills and Donna Sylvester
University of Washington – Brian Hopkins
Salve Regina University – William Stout

A number of individuals were helpful in the early stages of the project. In particular we would like to thank Duane Porter, University of Wyoming, for his suggestions on the mechanics of running workshops. We would also like to thank Marge Wechter, UMass Dartmouth, and Judith Leon for carefully reading the ATLAST proposal and for making helpful suggestions and Martha Kempe, UMass Dartmouth, for technical assistance with the grant proposal. And we thank Donald Albers of the Mathematical Association of America for his support of the project.

Richard Faulkenberry served as Assistant ATLAST Director and also assisted in coordinating the group activities at the San Diego Developers Workshop. Andre Weideman, Oregon State University, also assisted at the San Diego workshop. Steven Nash, Johns Hopkins University, Charles Johnson, College of William and Mary, and Joel Robbins, University of Wisconsin, contributed to the success of the workshops by appearing as guest speakers. Thanks also to Donald LaTorre, Clemson University, for his efforts in organizing sessions on innovations in teaching linear algebra at numerous conferences.

Special note should be taken of the services of Alexander Bondarenko, an undergraduate student at UMass Dartmouth. Alex worked through all of the exercises selected for this book. His comments and suggestions helped us to make significant improvements. Additionally, Alex developed a number of very nice ATLAST M-files that take full advantage of MATLAB's graphical user interface capabilities. Alex also helped to set up the ATLAST Web page.

We are also greatly indebted to Emily Moore of Grinnell University for her work on updating and improving the ATLAST M-files. Emily suggested and made significant improvements in many of our M-files. Her skills as a MATLAB programmer are quite impressive.

Finally, the editors would like to thank all of the ATLAST workshop participants. Their enthusiasm, energy and spirit have contributed so much to the project. They are the ones responsible for this book.

S. J. Leon, ATLAST Project Director

ATLAST Contributors

All of the ATLAST participants have been contributors in some way. A large percent of the participants from the 1992 and 1993 workshops submitted computer exercises for inclusion in the ATLAST collection. It's not possible to assign credit to every exercise or project individually, since many of these submissions were actually the result of collaborative efforts during workshops and since many people submitted similar exercises. For example, eight participants independently submitted similar projects on using matrices to send coded messages. Therefore we include a list of all participants.

All of the participants in the 1994 Developers Workshop deserve special recognition for substantial contributions to the present volume. Accordingly, we provide a separate listing of these individuals.

The participants in the two 1995 workshops and three 1996 workshops developed sets of lesson plans for teaching linear algebra. When the collection of lesson plans is complete, a selection of the best will be edited and made available to the general public through the ATLAST Web page (http://www.umassd.edu/SpecialPrograms/ATLAST). Selected lesson plans will also be included in future editions of this book. A list of the 1996 workshop participants will be included in the next edition of this book.

1994 ATLAST Developers Workshop

David W. Boyd, Valdosta State College
Thomas W. Cairns, The University of Tulsa
Ralph Czerwinski, Millikin University
John W. Davenport, Georgia Southern University
Luz Maria DeAlba, Drake University
Richard H. Elderkin, Pomona College
Larry C. Grove, University of Arizona
Thomas A. Hern, Bowling Green State University
Larry Neal, East Tennessee State University
George D. Poole, East Tennessee State University
Larry Riddle, Agnes Scott College
Lila F. Roberts, Georgia Southern University
Jeff Stuart, University of Southern Mississippi
Gary Thompson, Virginia Commonwealth University
Don Tosh, Evangel College
Avi Vardi, Drexel University
James R. Weaver, University of West Florida
Andre Weideman, Oregon State University
Jane Wells, Governors State University
Bruce Yoshiwara, Los Angeles Pierce College
T. J. Ypma, Western Washington University

ATLAST 1992 and 1993 Participants

Glenn Adamson, Ottawa University
Victor Akatsa, Chicago State Univ.
Margaret Allen-Gorlin, Middlesex County Coll.
Birgit Aquilonius, West Valley College
Donald F. Bailey, Trinity University
Joseph A. Ball, Virginia Polytechnic Inst.
Richard Barshinger, Penn State - Scranton
Linda Becerra, Univ. of Houston - Downtown
Karen Z. Benbury, Bowie State University
Fariba Bigdeli-Jahed, Kentucky State University
Carolyn L. Blaine, Christian Brothers University
Marilyn Blockus, San Jose State University
David W. Boyd, Valdosta State College
Eddy Joe Brackin, University of North Alabama
Shirley M. Branan, Birmingham-Southern College
Gary Brown, College of Saint Benedict
Cynthia Burnham, San Jose City College
Thomas W. Cairns, The University of Tulsa
Judith N. Cederberg, St. Olaf College

B. Carol Adjemian, Pepperdine University
Ricardo Alfaro, Univ. of Michigan, Flint
Steven M. Amgott, Widener University
Said Bagherieh, West Virginia Inst. of Tech.
V. K. Balakrishnan, University of Maine
Zeev Barel, Hendrix College
Sarah Bates, Calhoun Community College
Janet Beery, University of Redlands
Katalin A. Bencsath, Manhattan College
Ronald C. Biggers, Kennesaw State College
Patricia M. Blitch, Lander College
Nola Blye, Cheyney Univ. of Pennsylvania
Sylvia T. Bozeman, Spelman College
Sheryl Brady, SUNY Purchase
S. Allen Broughton, Cleveland State University
James T. Bruening, Southeast Missouri State U.
Zhixiong Cai, Barton College
Oresto N. Castillo, Alabama A&M University
Benito Chen, University of Wyoming

Mei-Qin Chen, The Citadel
Allan C. Cochran, Univ. of Arkansas at Fayetteville
Richard A. Cooper, Trinity University
Daniel W. Cunningham, SUNY Buffalo
Ellen Cunningham, Saint Mary-of-the-Woods Coll.
Stephen Curry, Spring Hill College
Ali A. Daddel, University of California, Davis
Jerome Dancis, University of Maryland
Karabi Datta, Northern Illinois University
Tilak de Alwis, Southeastern Louisiana University
Ralph DeMarr, University of New Mexico
David J. DeVries, Georgia College
Karen E. Donnelly, Saint Joseph's College
William W. Durand, Henderson State University
Nina Edelman, Spring Garden College
Thomas E. Elsner, GMI Eng. and Management Inst.
Mohamed El-Ansary, Cal State Bakersfield
Tom Falbo, Ohlone College
Francis G. Florey, Univ. of Wisconsin - Superior
Robert C. Forsythe, Carlow College
Mark Fuerniss, Northwestern Oklahoma State Univ.
Pratibha Ghatage, Cleveland State University
Mary Glaser, Tufts University
Edward H. Grossman, City College of NY
Carole L. Grover, Carlow College
Brad Gubser, Hiram College
Murli M. Gupta, George Washington Univ.
Richard Samuel Hall, Willamette University
P. R. Halmos, Santa Clara University
Robert Hanson, James Madison University
James L. Hartman, The College of Wooster
Maryam Hastings, Marymount College
Curtis Herink, Mercer University
Thomas A. Hern, Bowling Green State University
Tian-You Hu, Univ. of Wisconsin, Green Bay
Robert B. Hughes, Boise State University
Jane N. Ingram, Roanoke College
Ali A. Jafarian, University of New Haven
S. K. Jain, Ohio University
Clement Jeske, Univ. of Wisconsin, Platteville
Roy K. Johnson, Marian College of Fond du Lac
Wesley L. Jordan, Pace University, Westchester
Thomas W. Judson, University of Portland
Irvine J. Katz, George Washington University
John Kemper, University of St. Thomas
Gary Klatt, University of Wisconsin, Whitewater
Alan P. Knoerr, Occidental College
Lala B. Krishna, University of Akron
Ravinder Kumar, Alcorn State University
Steven H. Lameier, Thomas More College
Oksana Lassowsky, Albright College
Rebecca Lee, Bowie State University
Tan-Yu Lee, University of Alabama
Steve Ligh, Southeastern Louisiana University
Shinemin Lin, Bethel College

David T. Closky, College of Mount St. Joseph
Michael R. Colvin, California State Polytechnic U.
Ann Cox, Auburn University
Gladys Crates, Chattanooga State Tech. Comm. Coll.
John J. Currano, Roosevelt University
Ralph Czerwinski, Millikin University
Mel Damodaran, Univ. of Houston - Victoria
Nasser Dastrange, Buena Vista College
John W. Davenport, Georgia Southern University
Luz Maria DeAlba, Drake University
Telahun Desalegne, Florida Memorial College
Hung T. Dinh, Macalester College
Ida Doraiswamy, Elizabeth City State Univ.
Aniekan Ebiefung, Univ. of Tennessee, Chattanooga
Richard H. Elderkin, Pomona College
Ervin Eltze, Fort Hays State University
Michael A. Fahy, Chapman University
Sen Fan, Univ. of Minnesota - Morris
Jefferson Fong, The Master's College
Michael Frantz, University of La Verne
Chris J. Gardiner, Eastern Michigan University
Peter M. Gibson, Univ. of Alabama, Huntsville
Robert Grone, San Diego State University
Larry C. Grove, University of Arizona
Thomas Gruszka, Western New Mexico Univ.
Julio Guillen, Jersey City State College
Javad Habibi, Muskingum College
Jeremy Haefner, U. of Colorado at Colorado Springs
Arnold Hammel, Central Michigan University
Thomas Hanson, Northwestern State U. of Louisiana
Marian Harty, Edgewood College
Catherine Hayes, Mobile College
Mason Henderson, Oklahoma School of Sci. & Math.
Dan Hrozencik, Westminster College
David Hughes, Abilene Christian University
Kenneth E. Hummel, Trinity University
Robin Ives, Harvey Mudd College
Jay M. Jahangiri, Cal. State University, Bakersfield
Geraldine A. Jensen, Western Connecticut State Univ.
Kenneth Johnson, North Dakota State Univ., Fargo
Douglas Jones, McKendree College
Mahmoud Joudeh-Abu, Saint Paul's College
David Kaplan, York College of Pennsylvania
J. Kapoor, Anne Arundel Community College/UMBC
Gary S. Kersting, Sacramento City College
Robert C. Knapp, Univ. of Wisconsin, Whitewater
Daryl Kreiling, Univ. of Tennessee at Martin
Vatsala Krishnamani, Middle Tennessee State Univ.
Suda Kunyosying, Shepherd College
Robert W. Langer, Univ. of Wisconsin, Eau Claire
Cecelia Laurie, Univ. of Alabama, Tuscaloosa
Roger Lautzenheiser, Rose-Hulman Inst. of Tech.
Chi-Kwong Li, The College of William and Mary
En-Bing Lin, University of Illinois
Stephen L. Littell, Penn State University, Schuylkill

Bing Liu, College of St. Scholastica
B. A. Lotto, Lake Forest College
Bruce N. Lundberg, Grand Canyon University
Nachimuthu Manickam, DePauw University
Raj Markanda, Northern State University
John H. Mathews, California State U., Fullerton
John Maybee, University of Colorado, Boulder
Greg A. McClanahan, LaGrange College
Raymond McDaniel, Pembroke State University
Renate McLaughlin, Univ. of Michigan, Flint
David Meredith, San Francisco State University
George Miel, University of Nevada, Las Vegas
David F. Miller, Wright State Univ., Dayton
Hosien S. Moghadam, U. of Wisconsin, Oshkosh
John J. Morrell, Atlanta Metroplitan College
Kandasamy Muthuvel, U. of Wisconsin, Oshkosh
G. Naude, University of Pretoria
Donald A. Nelson, Middle Tennessee State Univ.
Steven Nimmo, Morningside College
Melvin A. Nyman, Alma College
John D. O'Neill, University of Detroit Mercy
Shivappa Palled, Methodist College
Sarah Patrick, Troy State University at Dothan
Marie E. Pink, Alverno College
George D. Poole, East Tennessee State University
Agnes T. Prindiville, McHenry County College
Thomas Putnam, Los Angeles Pierce College
Phil Quartararo Jr., Southern University
V. S. Ramamurthi, University of North Florida
Boris Reichstein, Catholic Univ. of America
Yuri Rojas-Ramirez, U. of Puerto Rico Mayaguez
Randy K. Ross, Morehead State University
Atul N. Roy, Culver-Stockton College
Sohindar S. Sachdev, Elizabeth City State Univ.
Ossama A. Saleh, U. of Tennessee, Chattanooga
Mark Sand, Northwest Missouri State University
K. P. Satagopan, Shaw University
Alice T. Schafer, Marymount University
Karen Schroeder, Bentley College
Daniel J. Scully, St. Cloud State University
Sally Sestini, Cerritos College
Mehrdad Simkani, University of Michigan, Flint
Muriel Skoug, Nebraska Wesleyan University
Ronald L. Smith, Univ. of Tennessee at Chattanooga
Steve Smith, Harding University
Ernie Solheid, California State Univ., Fullerton
Nesan S. Sriskanda, Allen University
Susan Staples, College of Staten Island - CUNY
Ralph Steinlage, University of Dayton
Tin-Yau Tam, Auburn University
Gary Thompson, Virginia Commonwealth Univ.
Tran Van Thuong, Missouri Southern State Coll.
Eve Torrence, Trinity College
Virginia Trasher, SUNY College at Geneseo
Vance W. Underhill, East Texas State University

Mark Lotspeich, Albertson College of Idaho
Sylvia C. Lu, University of Colorada at Denver
J. J. Malone, Worcester Polytechnic Institute
Shan Manickam, Western Carolina University
Paul B. Massell, The Johns Hopkins University
Carlton J. Maxson, Texas A&M University
Elizabeth Mayfield, Hood College
Otis B. McCowan, Belmont University
Cynthia McGinnis, Okaloosa Walton Comm. Coll.
Segundo D. Melendez, Univ. of Puerto Rico, Cayey
Dennis I. Merino, Southeastern Louisiana Univ.
Tom Miles, Central Michigan University
Janet E. Mills, Seattle University
Peter Monk, University of Delaware
Margaret Murray, Virginia Polytechnic Inst.
C. G. Naude, University of Pretoria
Larry Neal, East Tennessee State Univ.
Harry Newton, US Air Force Academy
Paula A. Norris, Delta State University
M. Lesley O'Connor, Queens College
James Paige, Wayne State College
Eddie C. Paramore, Tuskegee University
David Pelzl, Dr. Martin Luther College
Thomas W. Polaski, Winthrop University
Lanita Presson, Univ. of Alabama - Huntsville
Stephen K. Prothero, Willamette University
Wallace C. Pye, Univ. of Southern Mississippi
Kathryn M. Radloff, Ohio Wesleyan University
G. M. Reekie, Livingston University
Larry Riddle, Agnes Scott College
Lila F. Roberts, Georgia Southern University
Sharon Ross, Dekalb College
Andrzej Rusewicz, Hampden-Sydney College
Atma R. Sahu, Coppin State College
E. Salehi, Univ. of Nevada, Las Vegas
Sonja Sandberg, Radcliff College
Corinne Schaeffer, Mercyhurst College
Steven J. Schlicker, Grand Valley State Univ.
Paul Schuette, Georgia College
Brigitte Servatius, Worcester Polytechnic Inst.
Richard J. Shores, Lynchburg College
Balbir Singh, College of San Mateo
Jo E. Smith, GMI Eng. and Management Inst.
Scott A. Smith, Columbia College
James T. Snodgrass III, Xavier University
Selvaratnam Sridharma, Dillard University
J. Sriskandarajah, U. of Wisc. Center - Richland
Michael Stecher, Texas A&M University
Stanley L. Stephens, Anderson University
Gerald D. Taylor, Colorado State University
Samuel Thompson III, St. Mary's U. of San Antonio
Chris Tiahrt, Univ. of Nebraska at Lincoln
Don Tosh, Evangel College
Mary T. Treanor, Valparaiso University
Avi Vardi, Drexel University

Arnold H. Veldkamp, Dordt College
Martha L. Wallace, St. Olaf College
Rongdong Wang, Texas A&M University
Thomas Weber, Concordia College
Andre Weideman, Oregon State University
Paul H. Whitaker, Mount Vernon Nazarene Coll.
John R. Wicks, North Park College
June F. M. Winter, Marymount University
David H. Wood, University of Delaware
Henry L. Wyzinski, Indiana Univ., Northwest
Lisa M. Yates, Southern Oregon State College
Katherine Yoshiwara, Los Angeles Pierce College
Jianping Zhu, Mississippi State University

Diane M. Wagner, Regis University
Terry J. Walters, U. of Tenn. at Chattanooga
James R. Weaver, University of West Florida
Leben Wee, Montgomery College
Jane Wells, Governors State University
James S. White, Jacksonville State University
David E. Wilson, Wabash College
Roman Wong, Washington & Jefferson College
Gary L. Wood, Azusa Pacific University
Kemble Yates, Southern Oregon State College
Bruce Yoshiwara, Los Angeles Pierce College
T. J. Ypma, Western Washington University

ATLAST 1995 Participants

Olusola Akinyele, Bowie State University
Don Beken, Pembroke State University
William C. Calhoun, Kalamazoo College
Karen Clark, Trenton State College
Kevin Coates, Illinois Wesleyan University
Nasser Dastrange, Buena Vista College
Barbara J. Frank, St. Andrew's Presbyterian Coll.
Gary L. Ganske, Northwest Nazarene College
Angela Grant, Lincoln University
Mickey Holsey, Hampton University
Victor Klee, University of Washington
Alexander Koonce, University of Redlands
Richard Levin, Western Washington University
Xiaobo Liu, Clarkson University
V. P. Manglik, Elizabeth City State University
Sister Ann Mason OP., Aquinas College
Josiah Meyer, Elmira College
Leigh Ann Myers, NW State U. of Louisiana
Olympia Nicodemi, SUNY Geneseo
Robert L. Pour, Emory & Henry College
Leiba Rodman, The College of William and Mary
Patrick A. Rossi, Troy State University
Dan Seth, Morehead State University
Judy Seymour, Hood College
Kenneth Spackman, U. of N. Carolina Wilmington
Gordon Swain, Ashland University
Robert Underwood, Auburn U. at Montgomery
Tamsen Whitehead, Santa Clara University
Fred Worth, Henderson State University
Xueqi Zeng, Concordia College

Agnes Andreassian, Central College
Mikhail M. Bouniaev, Southern Utah University
Hasan A. Celik, California State Polytechnic U.
Charles C. Clever, South Dakota State Univ.
Ali A. Dad-del, Univ. of California, Davis
John Drew, The College of William and Mary
Mahmoud M. Fath El-Den, Fort Hays State U.
Martha Goss, Millsaps College
Philip E. Gustafson, Emporia State University
Prof. Dan Kalman, American University
Janusz Konieczny, Mary Washington College
Reginald Laursen, Luther College
Xiaoguang Li, Virginia Polytechnic Institute
Joe Mailhot, St. Martin's College
Dan Martinez, California State U., Long Beach
Jennifer McNulty, The University of Montana
John Mitchell, University of Alaska, Anchorage
Colm Mulcahy, Spelman College
Stephen Pierce, San Diego State University
Eugen Radian, College of Notre Dame
John F. Rossi, Virginia Polytechnic Institute
Jharna Sengupta, Elizabeth City State Univ.
Alicia Sevilla, Moravian College
Wasin So, Sam Houston State University
David Stanford, College of William and Mary
Bob Tilidetzke, Charleston Southern University
Ping Wang, Penn State University
Cynthia Woodburn, Pittsburg State University
Jeff Young, Virginia Polytechnic Institute
Marilyn A. Zopp, McHenry County College

Additional Contributors

Walter Gander, ETH Zurich

Deborah P. Levinson, Colorado College

CONTENTS

xxii

Chapter 1

Linear Systems

1.1   Exercises on Linear Systems

Linear Systems and Matrix Reduction

1. For each of the following linear systems of equations, reduce its augmented matrix step by step to row echelon form, and report the number of solutions for the linear system. Use as few steps as you can. Report your steps by listing the commands you used, and report the echelon form you got. (You may not use MATLAB's **rref** command, but you may use the ATLAST commands **rowcomb**, **rowswap**, and **rowscale**. You do not have to reduce the matrices to reduced row echelon form; any row echelon form will do.)

   (a)

$$\begin{aligned} -x_1 + 2x_2 + x_3 &= 0 \\ -2x_2 - x_3 &= 2 \\ -x_1 + x_2 \quad\;\; &= 3 \end{aligned}$$

   (b)

$$\begin{aligned} -x_2 - x_3 &= 1 \\ -x_1 \quad\;\; + x_3 &= 0 \\ x_1 - 2x_2 + x_3 &= -2 \\ x_1 - x_2 + x_3 &= 3 \end{aligned}$$

(c)

$$-x_1 + 3x_2 + 2x_3 = -3$$
$$-2x_1 + 6x_2 + 3x_3 = -1$$

(d)

$$3x_1 + 2x_2 + x_3 = 8$$
$$-2x_2 + 2x_3 = -2$$
$$-3x_1 - x_2 - 2x_3 = -7$$
$$3x_1 + x_2 + 2x_3 = 7$$

2. Construct a random 3×4 matrix A and a random 3×1 vector $\mathbf{b}$ by using the ATLAST commands:

```
A = randint(3,4,5)
b = randint(3,1,5)
```

Then compute the reduced row echelon forms of the coefficient matrix and the augmented matrix of the linear system $A\mathbf{x} = \mathbf{b}$ as follows:

```
rref(A)
rref([A,b])
```

(a) From the result of the command rref([A,b]), how can you tell whether the nonhomogeneous system $A\mathbf{x} = \mathbf{b}$ has a solution? If the system does not have a solution, replace the vector $\mathbf{b}$ by a new vector $\mathbf{b}$ so that the system does have a solution. Now use the echelon forms of A and $[A, \mathbf{b}]$ (with the new $\mathbf{b}$ if there is such) to explain why the homogeneous system $A\mathbf{x} = 0$ and the nonhomogeneous system $A\mathbf{x} = \mathbf{b}$ have the same number of solutions.

(b) Explain in general why $A\mathbf{x} = 0$ and $A\mathbf{x} = \mathbf{b}$ have the same number of solutions (assuming $A\mathbf{x} = \mathbf{b}$ has a solution).

3. (a) Find the general pattern for the reduced row echelon form of the *consecutive integers* matrix, consec(n). (Use rref to discover the pattern.)

(b) Derive this echelon form by hand for the general case. Hint: First subtract each row from the following row, starting at the bottom.

Rank of a Matrix

The *rank* of a matrix may be defined as the number of nonzero rows in any row echelon form of the matrix. You may use this definition in the next group of exercises.

4. (a) Make up by hand any 3 × 3 matrix of rank 2 that has no zero rows and in which no row is a multiple of another. Check the rank by using MATLAB's rref or rank command.

   (b) Make up by hand any 4 × 4 matrix of rank 3 that has no zero rows and in which no row is a multiple of another. Use MATLAB to check the rank.

   (c) Repeat (b) but choose the matrix so that no row is a linear combination of two others (i.e., no row can be written in the form $aR + bS$ where R and S are two other rows of the matrix and a and b are scalars).

5. In exercises (a) – (f) below, you are given a family of matrices $M_1, M_2, M_3, \ldots$, where M_n is $n \times n$. (See Appendix B to find out how to construct these matrices quickly.) Compute the rank of M_n for some small values of n until you see a pattern. Then (i) state the rank of M_n for all n; (ii) explain why this is indeed the correct rank of M_n for all n.

   (a) M_n = *checkerboard* matrix
   (b) M_n = *sign* matrix
   (c) M_n = *Jordan 0-block* matrix
   (d) M_n = *letter N* matrix
   (e) M_n = *letter X* matrix
   (f) $M_n = A_n + A_n^{\mathrm{T}}$, where A_n = *Jordan 0-block* matrix

Linear Systems in Action

6. In this exercise you will solve linear systems of equations to find sums of the form $\sum_{i=1}^{n} i^k$. You have probably seen a few sums of this form, such as

$$\sum_{i=1}^{n} i = \frac{n(n+1)}{2} = \frac{1}{2}n^2 + \frac{1}{2}n$$

$$\sum_{i=1}^{n} i^2 = \frac{n(n+1)(2n+1)}{6} = \frac{1}{3}n^3 + \frac{1}{2}n^2 + \frac{1}{6}n$$

One way to find the sum $\sum_{i=1}^{n} i^k$ is to use the fact (which we will not prove here) that the sum can be expressed as a polynomial of degree $k+1$ in n:

$$\sum_{i=1}^{n} i^k = a_{k+1}n^{k+1} + a_k n^k + \cdots + a_1 n + a_0 \qquad (1.1)$$

We can find the coefficients $a_{k+1}, \cdots, a_0$ by solving a linear system of $k + 2$ equations in these $k + 2$ unknowns; simply substitute the values $n = 0, 1, \cdots, k + 1$ in the above formula. For example, to find the sum $\sum_{i=1}^{n} i^2$, use the fact that $\sum_{i=1}^{n} i^2 = a_3 n^3 + a_2 n^2 + a_1 n + a_0$ and substitute the values $n = 0, 1, 2, 3$ into this equation. We thereby obtain the system of equations

$$
\begin{aligned}
0a_3 + 0a_2 + 0a_1 + a_0 &= 0 \\
a_3 + a_2 + a_1 + a_0 &= 1 \\
8a_3 + 4a_2 + 2a_1 + a_0 &= 5 \\
27a_3 + 9a_2 + 3a_1 + a_0 &= 14
\end{aligned}
$$

(a) Use MATLAB to solve the above system and thus confirm that

$$\sum_{i=1}^{n} i^2 = \frac{1}{3}n^3 + \frac{1}{2}n^2 + \frac{1}{6}n$$

(Hints: The coefficient matrix is a Vandermonde matrix and therefore can be constructed quickly using MATLAB's **vander** function. After solving the system, apply the **rats** function to your answer to convert it into a rational number.)

(b) Use MATLAB to find the sum $\sum_{i=1}^{n} i^4$.

7. In this exercise you will solve linear systems of equations to find magic squares. A *magic square* is a square matrix with integer entries in which all rows, all columns, and the two diagonals have the same sum. For example, the matrix

$$\begin{pmatrix} 8 & 1 & 6 \\ 3 & 5 & 7 \\ 4 & 9 & 2 \end{pmatrix} \tag{1.2}$$

is a magic square in which each row, each column, and each diagonal adds up to 15.

Suppose you are given a partially filled-in magic square. Can you fill in the rest? For example, is there a magic 3×3 square whose first row is [8 1 3]? We start by labeling all the unknown entries:

$$\begin{pmatrix} 8 & 1 & 3 \\ x_1 & x_2 & x_3 \\ x_4 & x_5 & x_6 \end{pmatrix}$$

Since the first row adds up to 12, so must the other two rows, the three columns, and the two diagonals:

$$x_1 + x_2 + x_3 = 12$$
$$x_4 + x_5 + x_6 = 12$$
$$8 + x_1 + x_4 = 12$$
$$1 + x_2 + x_5 = 12$$
$$3 + x_3 + x_6 = 12$$
$$8 + x_2 + x_6 = 12$$
$$3 + x_2 + x_4 = 12$$

(a) Solve the above system of seven nonhomogeneous equations in the six unknowns $x_1, \cdots, x_6$. To solve the system, apply rref to the augmented matrix, since we do not know in advance whether the system has no solutions, one solution, or infinitely many solutions. (In fact, you will find a unique solution, although not all of the x_i's will be positive.)

(b) For each of the partially filled-in magic squares below, use the above technique to fill in the rest of the matrix. Report whether there is no such magic square, exactly one, or infinitely many. Although it is sometimes required that the entries of a magic square be consecutive positive integers, as in the example (1.2), we ask only that they be integers.

$$\begin{pmatrix} 1 & ? & ? \\ ? & 2 & ? \\ ? & ? & 3 \end{pmatrix}, \begin{pmatrix} 1 & ? & 3 \\ ? & 2 & ? \\ ? & ? & ? \end{pmatrix}, \begin{pmatrix} 1 & ? & 3 \\ 2 & ? & ? \\ ? & ? & ? \end{pmatrix}, \begin{pmatrix} 1 & ? & ? \\ ? & 3 & 2 \\ 1 & ? & ? \end{pmatrix}$$

1.2 Projects on Linear Systems

1. Graphical Analysis of Linear Systems

Analyze each of the systems of linear equations below as follows. Each equation has the form $ax + by = c$, and the graph of such an equation is a line. Use the ATLAST command **plotline** to graph each of the equations in the linear system. Then, from the picture, state how many solutions the system has. If there is a unique solution, estimate it graphically; otherwise, describe the system geometrically and explain why there is not a unique solution. Here are the commands you will need for (a):

```
plotline(4,1,7)
hold on
plotline(3,−2,−3)
grid
hold off
```

(The **hold on** command freezes the current plot so that the second **plotline** command adds to the existing plot. The **grid** command displays a grid, which helps in estimating coordinates. The **hold off** command unfreezes the plot.)

(a)

$$4x + y = 7$$
$$3x - 2y = -3$$

(b)

$$4x + y = 7$$
$$-8x - 2y = -4$$

(c)

$$3x - 2y = -3$$
$$-6x + 4y = 6$$

(d)

$$4x + y = 7$$
$$x = 2$$

(e)

$$4x + y = 7$$
$$3x - 2y = -3$$
$$x + 3y = -1$$

2. Uniqueness of the Reduced Row Echelon Form

[This project requires teams.]

(a) Each team is required to reduce the matrix

$$\begin{pmatrix} 0 & -1 & 1 \\ -2 & -1 & 1 \\ 2 & -2 & 2 \end{pmatrix}$$

step by step to reduced row echelon form. You may not use the **rref** command except to check your result. Use instead the ATLAST commands **rowcomb**, **rowswap**, and **rowscale**, and record the commands you use. One team must start by swapping rows 1 and 2 and using the new $(1, 1)$ entry as the first pivot; the other team must start by swapping rows 1 and 3 and using the new $(1, 1)$ entry as the first pivot.

(b) Explain why the reduced row echelon forms produced by the two teams are identical to one another and to the matrix produced by the MATLAB command **rref(A)**.

3. Designing a Ski Jump

A convenient way to draw curves of a desired shape is to select some points on the curve and find a polynomial whose graph goes through these points. Two points, for example, determine a unique line, which is the graph of a polynomial of degree 1 (if the line is not vertical). Three noncollinear points determine a unique parabola, which is the graph of a polynomial of degree 2 (if the points have distinct x coordinates). Four points determine a unique polynomial of degree 3, and so on.

Suppose we want a curve through the points $(0, 7), (1, 6), (2, 9)$. We will find the unique quadratic polynomial

$$P(x) = ax^2 + bx + c$$

whose graph goes through these points. Substituting the values for x and y, we get a linear system:

$$0a + 0b + 1c = 7$$
$$1a + 1b + 1c = 6$$
$$4a + 2b + 1c = 9$$

(a) Solve for the unknown coefficients a, b, c by applying MATLAB's backslash operator \ to the system $A\mathbf{p} = \mathbf{y}$, where A is the coefficient matrix of the above system, $\mathbf{p}$ is the vector of unknowns $[a, b, c]^T$, and $\mathbf{y}$ is the column vector for the right side.

(b) Graph the polynomial $P(x) = ax^2 + bx + c$ whose coefficients a, b, c you found in (a), and check visually that its graph is a parabola which goes through the three given points. Use the following commands:

```
u = 0:0.1:2;
v = polyval(p,u);
plot(u,v)
```

(c) Find the coefficients of the cubic polynomial

$$P(x) = ax^3 + bx^2 + cx + d$$

whose graph goes through the points $(-2, 6), (1, 4), (2, 3), (3, -2)$. Hint: To speed up the construction of the coefficient matrix A, use the MAT-LAB commands

```
x = [-2;1;2;3]
A = vander(x)
```

(You might discover what **vander** does by examining the columns of A from right to left.)

(d) As in (b), check your answer to (c) visually; however, change the values of u to u = −2:.1:3;.

(e) Find the coefficients of the fourth-degree polynomial

$$P(x) = ax^4 + bx^3 + cx^2 + dx + e$$

whose graph goes through the points $(0, 0), (1, 1), (-1, 3)$, and whose slope at $x = -1$ is 20 and at $x = 1$ is 9. (This time **vander** will not be of use.)

(f) As in (b), check your answer to (e) visually; however, change the values of u to u = −1:0.01:1;.

(g) Design a ski jump that has the following specifications. The ski jump starts at a height of 100 feet and finishes at a height of 10 feet. From start to finish, the ski jump covers a horizontal distance of 120 feet. A skier using the jump will start off horizontally and will fly off the end at a 30 degree angle from the horizontal. Find a polynomial whose graph is a side view of the ski jump.

(h) As in (b), check your answer to (g) visually.

Note: The polynomials in parts (a) and (c) above are called *Lagrange polynomials*. These are polynomials that go through a given set of points but whose derivatives are not specified (as they were in parts (e) and (g)).

4. Designing With Cubic Splines

As in the preceding project, we seek a smooth curve that goes that goes through some given data points. It is not always wise to use a single polynomial, as we did in that project. If the number of data points is large, the polynomial will have a high degree and hence will have a great many maximums and minimums, which may yield a much bumpier curve than we want.

A better choice of curve may be a *cubic spline*, in which we connect consecutive points with a cubic (i.e., degree 3) polynomial and where each interval uses a different cubic. Here is a graph of a cubic spline through eight data points whose x coordinates are $x = 1, 2, \ldots, 8$.

This spline uses seven cubics: s_1 is defined on $1 \le x \le 2$, s_2 is defined on $2 \le x \le 3$, and so on. The graph looks smooth because we require that adjacent cubics have matching first and second derivatives. For example, at $x = 2$ we require that $s_1'(2) = s_2'(2)$ and $s_1''(2) = s_2''(2)$ in addition to requiring that $s_1(2) = y_2$ and $s_2(2) = y_2$.

(a) Plot the data points $(-1, 4)$, $(0, 5)$, and $(1, 2)$ by using the following MATLAB commands:

```
x = [−1 0 1]
y = [4 5 2]
plot(x,y,'x')
hold on
```

(b) Derive the cubic spline for these data points as follows. There are two cubics:

$$s_1(x) = a_1x^3 + a_2x^2 + a_3x + a_4, \quad -1 \le x \le 0$$
$$s_2(x) = a_5x^3 + a_6x^2 + a_7x + a_8, \quad 0 \le x \le 1$$

Find 6 equations in the 8 unknowns $a_1, \dots, a_8$ by using

$$s_1(-1) = 4$$
$$s_1(0) = 5$$
$$s_2(0) = 5$$
$$s_2(1) = 2$$
$$s_1'(0) = s_2'(0)$$
$$s_1''(0) = s_2''(0)$$

We need two more equations to determine the unknowns uniquely. One way this is often done is to require that the second derivatives at the two endpoints are both zero. In our example, this means

$$s_1''(-1) = 0$$
$$s_2''(1) = 0$$

Solve the above 8×8 system of equations using MATLAB's backslash operator \.

Note: There are much more efficient ways to organize the computation of the cubic spline polynomials. You might find this subject discussed in more detail in your textbook, or you might ask your instructor for more information.

(c) Graph the cubic spline you found in (b) as follows:

```
x1 = −1:0.1:0;
s1 = a1*x1.^3 + a2*x1.^2 + a3*x1 + a4;
plot(x1,s1)
hold on
x2 = 0:0.1:1;
s2 = a5*x2.^3 + a6*x2.^2 + a7*x2 + a8;
plot(x2,s2)
```

MATLAB has a built-in command **spline** for computing a cubic spline. In the next two exercises, you may use this shortcut.

(d) Graph the cubic spline that goes through eight given points. Choose evenly spaced x coordinates and eight random y coordinates. Here is one method:

```
x = 1:8
y = rand(1,8)
u = 1:0.1:8;
v = spline(x,y,u);
plot(x,y,'*',u,v)
```

(e) Graph a cubic spline to fit a picture of a real object. Here is a suggested procedure. Find a picture of an object whose profile is not too bumpy; select a portion of the profile curve that can be seen as the graph of a function; pick no more than eight points on this portion of the curve; assign coordinates to these points; and, finally, graph the cubic spline through these points, as in (d). If your graph does not match the original profile curve, explain why.

5. Flow in a Network

The picture below represents a network of interconnected water pipes, with the direction of flow indicated by arrows.

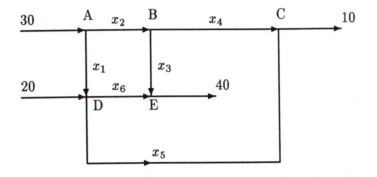

Suppose it is necessary to repair the pipe between connections D and E, and that we therefore want the flow in this section of pipe to be as small as possible. The flow at the connecting points A, B, C, D, E is regulated by valves, so we can control the flows $x_1, x_2, x_3, x_4, x_5, x_6$. (All the flows are measured in ft^3 per minute.)

Of course the amount of water entering a connecting point in one minute must equal the amount leaving. This leads to five equations in the six unknown flows. For example, conservation of flow at connecting point A yields the equation $x_1 + x_2 = 30$.

(a) Write down the remaining four equations and check that the augmented matrix for the system of five equations is

$$\begin{pmatrix} 1 & 1 & 0 & 0 & 0 & 0 & 30 \\ 0 & -1 & 1 & 1 & 0 & 0 & 0 \\ 0 & 0 & 0 & 1 & 1 & 0 & 10 \\ -1 & 0 & 0 & 0 & 1 & 1 & 20 \\ 0 & 0 & 1 & 0 & 0 & 1 & 40 \end{pmatrix}$$

As we will see, the system of equations has infinitely many solutions. However, not all these solutions make sense; since we cannot reverse any of the flows, we require that all the x_i's are non-negative. Such a solution is said to be *feasible*.

(b) From the picture alone, guess at a feasible solution of the system of equations. (Trial and error should work, since there are lots of feasible solutions.) Check your solution by plugging it into the system of equations in (a).

(c) Find all the solutions of the system of equations in (a). For example, you could apply **rref** to the augmented matrix and then use the echelon form to find all the solutions by hand. (There are two free variables; set up your solution so that the free variables are x_5 and x_6.)

(d) Since each x_i must be non-negative, your answer to (c) yields six inequalities, $x_1 \geq 0, \ldots, x_6 \geq 0$. Express each of these inequalities as inequalities for x_5, x_6, or $x_5 + x_6$.

(e) Recall that we want the flow in the section of pipe from D to E to be as small as possible. In other words, we want to minimize x_6. Use the inequalities in (d) to find the smallest value of x_6 for which there is a feasible solution.

(f) Write the six flows you found in (e) on the picture of the network of pipes, and check that conservation of flow holds.

(g) Now reverse the direction of flow in the pipe from B to E. In other words, change the inequality $x_3 \geq 0$ to $x_3 \leq 0$. Find the minimum value of x_6 in this new situation. Describe how your answer here differs from your answer to (e).

Note: This type of problem can be solved most efficiently by the methods of *linear programming*. You might find this subject discussed in your textbook, or you might ask your instructor for information about it.

Chapter 2

Matrix Algebra

2.1 Exercises on Matrix Algebra

Rules of Algebra

1. The following rules of algebra for matrices look like rules of algebra for real numbers. However, most of them are false for matrices. Use MATLAB to find out which rules are false, and, for each false rule, give a "counter-example" (an example of matrices for which the rule is false). To test these rules, you might use the ATLAST command **randint**, as in **A = randint(3,3,5)**, which creates random 3×3 matrices with integer entries between -5 and 5. For variety, also create some matrices that are all zero except for one entry.

   (a) $A + B = B + A$
   (b) $AB = BA$
   (c) If $AB = O$ then $A = O$ or $B = O$
   (d) If $A^2 = O$ then $A = O$
   (e) $(A + B)^2 = A^2 + 2AB + B^2$
   (f) $(A - B)(A + B) = A^2 - B^2$
   (g) $A(B + C) = AB + AC$
   (h) $(A + B)C = CA + CB$
   (i) $(AB)^2 = A^2 B^2$

Matrices in Action

2. A furniture store stocks three sizes of beanbag chairs — small, medium, and large. The sales for the months of September, October, and November were $S1 = [25\ 20\ 19]$, $S2 = [22\ 31\ 34]$, and $S3 = [19\ 33\ 26]$, respectively. (The first, second, and third entries of each vector represent the sales of small, medium, and large chairs, respectively.)

Enter $S1, S2, S3$ as the rows of a 3×3 matrix S in MATLAB. Answer each of the following questions by using MATLAB commands, and state the commands you used.

   (a) How many of each type of beanbag chair were sold during the three-month period?
   (b) What was the average number of sales per month for each type of beanbag chair?
   (c) If the quantity of stock on September 1 was $q = [120\ 95\ 87]$ and no new shipment was received during the three-month period, how many of each type of beanbag chair were left on December 1?
   (d) If the price in dollars for the three types of beanbag chairs was $[23\ 28\ 35]$, what was the total sales in dollars for each month? (Hint: Enter the price vector as a single column.)
   (e) What was the total sales in dollars for the large beanbag chairs over the three-month period?

3. In an experiment studying the effects of various diets on apes, two types of apes were studied and each type was divided into two age groups. The number of apes of each type and age is given by the 2×2 matrix A:

$$\begin{array}{cc} \text{young} & \text{old} \end{array}$$
$$\left(\begin{array}{cc} 70 & 50 \\ 30 & 60 \end{array}\right) \begin{array}{l} \text{chimpanzees} \\ \text{gibbons} \end{array}$$

The number of grams of protein, carbohydrates, and fat ingested per day by each young ape and each old ape is given by the 2×3 matrix B:

$$\begin{array}{ccc} \text{prot.} & \text{carbo.} & \text{fat} \end{array}$$
$$\left(\begin{array}{ccc} 25 & 40 & 20 \\ 15 & 30 & 25 \end{array}\right) \begin{array}{l} \text{young} \\ \text{old} \end{array}$$

Compute the matrix product AB, and state the meaning of each entry in the product.

Special Matrices

4. In exercises (a) – (i) below, you are given a family of $n \times n$ matrices $M_1, M_2, M_3, \cdots$. (See Appendix A to find out how to construct these matrices quickly.) Compute M_n^2 for some small values of n until you see a pattern. Then (i) describe the pattern for M_n^2 for all n (either in words or by writing out M_n^2 in general, using ... where appropriate); (ii) by looking closely at the entries of M_n and at how the square of a matrix is computed, explain some part of this pattern (for example, by explaining one entry of M_n^2).

   (a) $M_n = J_{nn} = $ *ones* matrix
   (b) $M_n = G_n = $ *grid* matrix
   (c) $M_n = C_n = $ *checkerboard* matrix
   (d) $M_n = $ *sign* matrix
   (e) $M_n = $ *letter L* matrix
   (f) $M_n = $ *letter N* matrix
   (g) $M_n = $ *letter Z* matrix
   (h) $M_n = $ *letter X* matrix
   (i) $M_n = $ *letter H* matrix (which is defined only for odd n)

5. In exercises (a) – (g) below, you are given a family of $n \times n$ matrices $M_1, M_2, M_3, \cdots$. (See Appendix A to find out how to construct these matrices quickly.) Compute M_n^k for some small values of n and k until you see a pattern. Then (i) describe the pattern for M_n^k for all n and k, where $k = 1, 2, 3, \cdots$ (either in words or by writing out M_n^k in general, using ... where appropriate); (ii) by looking closely at the entries of M_n and at how the powers of a matrix are computed, explain some part of this pattern (for example, by explaining one entry of M_n^k).

   (a) $M_n = $ *sign* matrix
   (b) $M_n = B_n = $ *backwards identity* matrix
   (c) $M_n = $ *Jordan 0–block* matrix
   (d) $M_n = $ *cyclic* matrix
   (e) $M_n = $ *letter N* matrix

(f) $M_n = $ *letter Z* matrix

(g) $M_n = $ *letter X* matrix

6. A square matrix is *upper triangular* if its entries below the main diagonal
are 0. Use the MATLAB command triu(rand(5)) to construct some 5×5
random upper-triangular matrices; multiply them together. What do you
observe about the form of the product of two upper-triangular matrices?
State this conclusion as a theorem and explain how one might prove it.

7. A square matrix is *strictly upper triangular* if its entries on and below the
main diagonal are 0. Use the command triu(rand(5),1) to construct some
5×5 random strictly upper-triangular matrices; multiply them together.

   (a) What do you observe about the form of the product of two strictly
   upper-triangular matrices? State this conclusion as a theorem and
   explain how one might prove it.

   (b) Compute the powers A^k of a strictly upper-triangular matrix A for
   $k = 2, 3, 4, 5$. State what you observe about the form of A^k as a
   theorem.

8. In this exercise, you will explore properties of the $n \times n$ *cyclic* matrix E.

   (a) Generate the 5×5 *cyclic* matrix

   $$E = \begin{pmatrix} 0 & 1 & 0 & 0 & 0 \\ 0 & 0 & 1 & 0 & 0 \\ 0 & 0 & 0 & 1 & 0 \\ 0 & 0 & 0 & 0 & 1 \\ 1 & 0 & 0 & 0 & 0 \end{pmatrix}$$

   Also generate several random 5×5 matrices A (for example, by using
   A = randint(5,5,9)). Compute the products EA and AE, and describe
   the effect on A. State your conclusion as a general theorem about the
   result of multiplying any $n \times n$ matrix on the right or left by the $n \times n$
   cyclic matrix.

   (b) Use your theorem in (a) to predict the effect on any $n \times n$ matrix A
   of the products $E^k A$ and AE^k, for $k = 2, \cdots, n-1$. Then check your
   conclusion by using MATLAB to compute $E^k A$ and $A^k E$ for several
   values of n and k and for several matrices A.

Linear Systems and Matrix Algebra

9. Consider the matrix

$$A = \begin{pmatrix} 7 & 6 \\ 2 & 3 \end{pmatrix}$$

We wish to determine the set $\mathcal{V}$ of all 2×2 matrices B that commute with A, i.e., for which $AB = BA$. For example, the identity matrix I and the matrix A itself are in $\mathcal{V}$, since $AI = IA$ and $AA = AA$.

(a) To find all the matrices B that commute with A, first write B with unknown entries:

$$B = \begin{pmatrix} x & y \\ u & v \end{pmatrix}$$

Multiply out AB and BA, set corresponding entries equal to each other, and obtain thereby a system of four linear equations in the four unknowns x, y, u, and v.

(b) Solve the system obtained in (a). Use the solution to write out the general form of matrices in $\mathcal{V}$.

(c) Show how to obtain I and A from the general form you derived in (b).

(d) Check that the following matrix commutes with A:

$$C = \begin{pmatrix} 4 & 9 \\ 3 & -2 \end{pmatrix}$$

Therefore C is in $\mathcal{V}$. Show how to obtain C from the general form derived in (b).

10. Form a random 3×4 matrix of rank 2 by using the ATLAST command A = randint(3,4,5,2). Compute its reduced row echelon form by using the MATLAB command rref(A).

(a) Using the above echelon form, find, by hand, two nonzero solutions $\mathbf{x}_1$ and $\mathbf{x}_2$ of $A\mathbf{x} = 0$ such that neither solution is a multiple of the other. Choosing scalars a and b at random, check that $a\mathbf{x}_1 + b\mathbf{x}_2$ is also a solution of $A\mathbf{x} = 0$; explain how you checked this. Formulate a corresponding general statement about all homogeneous linear systems.

(b) Choose a random 4×1 vector **p** and compute **b** = A*p. (So **p** is a solution of the nonhomogeneous system $A\mathbf{x} = \mathbf{b}$.) Using $a, b, \mathbf{x}_1$, and $\mathbf{x}_2$ from (a), check that $\mathbf{p} + a\mathbf{x}_1 + b\mathbf{x}_2$ is also a solution of $A\mathbf{x} = \mathbf{b}$; explain how you checked this. Formulate a corresponding general statement about all nonhomogeneous linear systems.

11. In this exercise, you will use systems of equations to better understand the rules of matrix algebra. Create three matrices A, B, and C as follows:

```
A = randint(3,3,5,3)
B = randint(3,1,5)
C = randint(1,3,5)
```

(a) The equation $AX = B$ is a system of three equations in three unknowns that can be solved using MATLAB's backslash operator: X = A\B. Do so.

(b) Now look at $AX = B$ as an equation that can be solved by using A^{-1} and rules of algebra. Find such a method, use MATLAB to compute the solution X by this method, and check that this answer agrees with your answer in (a).

(c) The equation $YA = C$ is also a system of three equations in three unknowns; it can be solved using MATLAB's slash operator: Y = C/A. Do so.

(d) As in (b), solve $YA = C$ by using rules of algebra and check that this answer agrees with your answer in (c).

(e) Solve the equation $YA = C$ a third way: convert it to the form $AX = B$ by using transposes, and solve it by your method in (b). Check that this answer also agrees with your answer in (c).

12. (a) Construct a random 4×3 matrix by using the ATLAST command A = randint(4,3,5). Then compute C = rref([A,eye(4)]). Find a nonsingular submatrix B of C with the property that BA is the reduced row echelon form of A.

(b) More generally, if A is any matrix and if the augmented matrix $[A \ I]$ is row equivalent to a matrix C, explain how to find a submatrix B of C with the following property: BA is row equivalent to A and equals the left part of C.

The Inverse of a Matrix

13. In this exercise, you will discover the general formula for the inverse of any 2×2 matrix and conditions under which the inverse exists.

(a) Suppose B is the inverse of A, where

$$A = \begin{pmatrix} a & b \\ c & d \end{pmatrix} \quad \text{and} \quad B = \begin{pmatrix} w & x \\ y & z \end{pmatrix}$$

Show that

$$aw + 0x + by + 0z = 1$$
$$0w + ax + 0y + bz = 0$$
$$cw + 0x + dy + 0z = 0$$
$$0w + cx + 0y + dz = 1$$

(b) Use MATLAB's Symbolic Toolbox to solve the above linear system of four equations for the unknowns w, x, y, z.

(c) From your answer to (b), find a necessary and sufficient condition for A to have an inverse.

(d) Verify by hand that your answer to (b) is correct. (Hint: You do not need to find A^{-1} by hand; since you have a formula for A^{-1}, there is an easy multiplication you can carry out to verify that it is correct.)

14. This exercise focuses on the definition of the matrix inverse, which you might want to review now. Note that the inverse is defined indirectly; rather than being defined by a formula that tells you how to compute the inverse, it is defined indirectly by equations that the inverse and the original matrix must satisfy. So in this exercise, you will only need to verify these equations, which will be a simple hand computation.

(a) Compute BA and AB for the matrices

$$A = \begin{pmatrix} 3 & 2 \\ 4 & 3 \end{pmatrix} \quad \text{and} \quad B = \begin{pmatrix} 3 & -2 \\ -4 & 3 \end{pmatrix}$$

Does it follow from these computations that A is nonsingular and that $B = A^{-1}$? Does it also follow that B is nonsingular and $A = B^{-1}$? Explain, using the definition of matrix inverse.

(b) Repeat part (a) for the matrices

$$A = \begin{pmatrix} 1 & 0 & 1 \\ 1 & 1 & 0 \\ -1 & 2 & -2 \end{pmatrix} \quad \text{and} \quad B = \begin{pmatrix} -2 & 2 & -1 \\ 2 & -1 & 1 \\ 3 & -2 & 1 \end{pmatrix}$$

(c) Repeat part (a) for the permutation matrices

$$A = \begin{pmatrix} 0 & 0 & 0 & 0 & 1 \\ 1 & 0 & 0 & 0 & 0 \\ 0 & 1 & 0 & 0 & 0 \\ 0 & 0 & 0 & 1 & 0 \\ 0 & 0 & 1 & 0 & 0 \end{pmatrix} \quad \text{and} \quad B = \begin{pmatrix} 0 & 1 & 0 & 0 & 0 \\ 0 & 0 & 1 & 0 & 0 \\ 0 & 0 & 0 & 0 & 1 \\ 0 & 0 & 0 & 1 & 0 \\ 1 & 0 & 0 & 0 & 0 \end{pmatrix}$$

(d) For the matrices $A = \begin{pmatrix} 1 & 1 & 1 \\ 1 & 1 & 0 \end{pmatrix}$ and $B = \begin{pmatrix} 1 & 1 \\ -1 & 0 \\ 1 & -1 \end{pmatrix}$, compute

AB. Does it follow from this computation that A is nonsingular and $B = A^{-1}$? Explain.

15. This exercise asks you to consider a variety of methods for computing a matrix inverse.

(a) Generate a random nonsingular 4×4 matrix A by the ATLAST command $A = \text{randint(4,4,1,4)}$. Compute A^{-1} by MATLAB's inv function.

(b) Construct the four standard basis vectors in $\mathbf{R}^4$:

$$\mathbf{e}_1 = \begin{pmatrix} 1 \\ 0 \\ 0 \\ 0 \end{pmatrix}, \mathbf{e}_2 = \begin{pmatrix} 0 \\ 1 \\ 0 \\ 0 \end{pmatrix}, \mathbf{e}_3 = \begin{pmatrix} 0 \\ 0 \\ 1 \\ 0 \end{pmatrix}, \mathbf{e}_4 = \begin{pmatrix} 0 \\ 0 \\ 0 \\ 1 \end{pmatrix}$$

and solve $A\mathbf{x}_i = \mathbf{e}_i$ for $\mathbf{x}_i$, where $i = 1, 2, 3, 4$. Explain how to find A^{-1} from your four solutions, and explain why this works.

(c) Augment the matrix A by the identity matrix, using the command

$$\text{AUG} = [\text{A, eye(4)}]$$

Find the row reduced echelon form of the augmented matrix, using rref(AUG). Explain how to find A^{-1} from the reduced matrix, and explain why this works.

16. *Input/output* models are used by economists to study the complex inter-relationships among elements of the economy. Here we examine a highly simplified input/output model. (A real one might have thousands of unknowns and equations.)

The *consumption matrix* tells how many units of each input (such as chemicals, food, and oil) are needed to produce one unit of output (also chemicals, food, and oil). Let the number of units of input be given by the vector

$$\mathbf{p} = \begin{pmatrix} p_1 \\ p_2 \\ p_3 \end{pmatrix} = \begin{pmatrix} \text{units of chemicals} \\ \text{units of food} \\ \text{units of oil} \end{pmatrix}$$

Suppose we know that the three outputs are given by the product

$$\mathbf{q} = \begin{pmatrix} \text{units of chemicals} \\ \text{units of food} \\ \text{units of oil} \end{pmatrix} = \begin{pmatrix} 0.2 & 0.3 & 0.4 \\ 0.4 & 0.4 & 0.1 \\ 0.5 & 0.1 & 0.3 \end{pmatrix} \begin{pmatrix} p_1 \\ p_2 \\ p_3 \end{pmatrix}$$

This says, for example, that inputs of p_1 units of chemicals, p_2 units of food, and p_3 units of oil yield $0.2p_1 + 0.3p_2 + 0.4p_3$ units of chemicals. We will write this equation more compactly as $\mathbf{q} = A\mathbf{p}$, where A is the above 3×3 matrix of numbers.

Now suppose there is demand for y_1 units of chemicals, y_2 units of food, and y_3 units of oil. Some of the inputs need to go to production, and only the remainder is available to satisfy demand. Since $A\mathbf{p}$ is consumed by production, only $\mathbf{p} - A\mathbf{p}$ is available to satisfy the demand $\mathbf{y}$. So to find out if the economy can satisfy the demand, we must solve

$$(I - A)\mathbf{p} = \mathbf{y}$$

for $\mathbf{p}$.

(a) Find out if the equation $(I - A)\mathbf{p} = \mathbf{y}$ has a solution for every possible demand vector $\mathbf{y}$. Hint: Not only must $I - A$ be nonsingular, but the solution $\mathbf{p}$ must have all non-negative entries. If $(I - A)^{-1}$ exists and has all non-negative entries, then $\mathbf{p} = (I - A)^{-1}\mathbf{y}$ will exist and have all non-negative entries.

(b) Change the $(1, 1)$ entry of A to 4.5. Why is A no longer a realistic consumption matrix?

Inverses of Special Matrices

17. In exercises (a) – (d) below, you are given a family of $n \times n$ matrices $M_1, M_2, M_3, \cdots$. (See Appendix A to find out how to construct these matrices quickly.) Compute M_n^{-1} for some small values of n until you see a pattern. Then (i) describe the pattern for M_n^{-1} for all n (either in words or by writing out M_n^{-1} in general, using $\ldots$ where appropriate); (ii) by using the definition of the inverse of a matrix, explain some part of this pattern (for example, by checking that one row or column of M_n^{-1} is correct).

(a) $M_n = $ *maximum* matrix

(b) $M_n = $ *minimum* matrix

(c) $M_n = $ *Jordan 1-block* matrix

(d) $M_n = $ *cyclic* matrix

18. Consider the 5×5 matrix

$$A = \begin{pmatrix} 0 & 1 & 0 & 0 & 0 \\ 0 & 0 & 1 & 0 & 0 \\ 0 & 0 & 0 & 1 & 0 \\ 0 & 0 & 0 & 0 & 1 \\ x & 0 & 0 & 0 & 0 \end{pmatrix}$$

for all possible values of the real number x.

(a) For what values of x is A nonsingular? Prove that your answer is correct.

(b) Use MATLAB to compute A^{-1} for several values of x. Then guess at the general formula for A^{-1}, and prove that your formula is correct.

(c) Use MATLAB to compute A^5 for several values of x. Then guess at the general formula for A^5, and use your answer to find infinitely many 5th roots of the 5×5 identity matrix.

(d) There is nothing special about 5×5 matrices in this problem. Based on your answer to (c), guess at a formula that gives infinitely many nth roots of the $n \times n$ identity matrix. Then check your formula by using MATLAB to compute the nth power of several such matrices for values of n other than 5.

Rank of a Matrix

The *rank* of a matrix may be defined as the number of nonzero rows in any row echelon form of the matrix. You may use this definition in the next group of exercises.

19. Construct several matrices by ATLAST commands of the form $A =$ randint(m,n,5,r), and vary m, n, and r. (This command generates a $m \times n$ matrix of rank r with entries between -5 and 5.) For each such matrix A, compute the ranks of A, A^T, AA^T, and A^TA. Guess at a relationship between the ranks of A, A^T, AA^T, and A^TA.

20. Construct several pairs of matrices A and B by repeated use of the ATLAST commands

    A = randint(m,n,5,r)
    B = randint(n,m,5,s)

    and vary m, n, r, and s.

    (a) For each pair of matrices A and B, record the ranks of A, B, AB, and BA:

	1	2	3	4	5	6	7	8	9	10
rank of A										
rank of B										
rank of AB										
rank of BA										

    (b) From your data in (a), guess at a relationship between the ranks of A, B, AB, and BA.

2.2   Projects on Matrix Algebra

1. Perspectives on Matrix Multiplication

The purpose of this project is to explore a variety of ways of thinking about and expressing the product of two matrices.

(a) Construct a random 3×4 matrix A and a random 4×5 matrix B by using the ATLAST commands

$$A = \mathsf{randint}(3,4,5)$$
$$B = \mathsf{randint}(4,5,5)$$

Compute the product AB. Also, multiply A by each column of B by using the commands

$$\mathsf{A*B(:,1),\ A*B(:,2),\ A*B(:,3),\ A*B(:,4),\ A*B(:,5)}$$

How are these five products related to the product AB? Describe, in general, how to view the product of two compatible matrices in terms of the product of the entire first matrix and the columns of the second matrix.

(b) For the same matrices A and B as in (a), multiply each row of A by the entire matrix B by using the commands

$$\mathsf{A(1,:)*B,\ A(2,:)*B,\ A(3,:)*B}$$

How are these three products related to the product AB? Describe, in general, how to view the product of two compatible matrices in terms of the product of the rows of the first matrix and the entire second matrix.

(c) Construct a random column 4-tuple $\mathbf{x}$ by using the command

$$\mathsf{x = randint(4,1,5)}$$

Compute the product $A\mathbf{x}$, where A is the matrix you constructed in (a). Also compute the following linear combination of the columns of A:

$$\mathsf{x(1)*A(:,1)+x(2)*A(:,2)+x(3)*A(:,3)+x(4)*A(:,4)}$$

How is this linear combination related to the product Ax? Describe, in general, how to view the product of two compatible matrices, in which the second has just one column, in terms of the columns of the first matrix.

(d) Construct a random row 4-tuple **y** by using the command

$$y = \text{randint}(1,4,5)$$

Compute the product $\mathbf{y}B$, where B is the matrix you constructed in (a). Also compute the following linear combination of the rows of B:

$$y(1)*B(1,:)+y(2)*B(2,:)+y(3)*B(3,:)+y(4)*B(4,:)$$

How is this linear combination related to the product $\mathbf{y}B$? Describe, in general, how to view the product of two compatible matrices, in which the first has just one row, in terms of the rows of the second matrix.

(e) Use your conclusions in (a) and (c) to answer the following question. Describe, in general, how to view the product of two compatible matrices in terms of the columns of the first matrix and the entries of the second matrix.

(f) Use your conclusions in (b) and (d) to answer the following question. Describe, in general, how to view the product of two compatible matrices in terms of the rows of the second matrix and the entries of the first matrix.

2. Symmetric and Skew–Symmetric Matrices

A matrix A is *symmetric* if $A^{\mathrm{T}} = A$. A matrix A is *skew–symmetric* if $A^{\mathrm{T}} = -A$. The purpose of this project is to discover properties of symmetric and skew–symmetric matrices. We can use MATLAB to construct random symmetric and skew–symmetric matrices as follows. If B is any square matrix, then $B + B^{\mathrm{T}}$ is symmetric and $B - B^{\mathrm{T}}$ is skew–symmetric (see (a) and (b) below). So we can construct random symmetric and skew–symmetric matrices by using the ATLAST commands

$$B = \text{randint}(n); \ A = B+B'$$

and

$$B = \text{randint(n)}; \ A = B-B'$$

respectively.

(a) Show that if B is any square matrix, then $B + B^T$ is symmetric.
(b) Show that if B is any square matrix, then $B - B^T$ is skew–symmetric.
(c) Generate some random 3×3 and 4×4 symmetric matrices A by the method shown above. What relationship exists between the rows of A and the columns of A? (This observation will give you a quick way to see visually whether a matrix is symmetric.)
(d) Generate some random 3×3 and 4×4 skew–symmetric matrices A by the method shown above. What relationship exists between the rows of A and the columns of A? What is special about the diagonal entries of A?
(e) Generate several random 3×3 and 4×4 symmetric matrices, and use them to explore the following possible properties of symmetric matrices. If you believe the property is true, prove it by using properties of the transpose. If you find the property is not true, give a "counter–example" (an example of matrices for which the property is false).

   i. If A and B are symmetric, is $A + B$ symmetric?
   ii. If A is symmetric and c is a scalar, is cA symmetric?
   iii. If A and B are symmetric, is AB symmetric?
   iv. If A is symmetric, is A^T symmetric?
   v. If A is symmetric and invertible, is A^{-1} symmetric?

(f) Generate several random 3×3 and 4×4 skew–symmetric matrices, and use them to explore the following possible properties of skew–symmetric matrices. If you believe the property is true, prove it by using properties of the transpose. If you find the property is not true, give a counter–example.

   i. If A and B are skew–symmetric, is $A + B$ skew–symmetric?
   ii. If A is skew–symmetric and c is a scalar, is the matrix cA skew–symmetric?
   iii. If A and B are skew–symmetric, is AB skew–symmetric?
   iv. If A is skew–symmetric, is A^T skew–symmetric?

v. If A is skew–symmetric and invertible, is the matrix A^{-1} skew–symmetric? (Use only 4×4 matrices for this part, as 3×3 skew–symmetric matrices are always singular.)

(g) If A and B are both $n \times n$ symmetric matrices and $AB = BA$, then AB is also symmetric. Use the properties of the transpose to prove this.

(h) If A and B are both $n \times n$ skew–symmetric matrices and $AB = BA$, what can you conclude about the product AB?

(i) If A is any square matrix, prove that A can be written as the sum of a symmetric matrix and a skew–symmetric matrix; that is, $A = B + C$ where B is symmetric and C is skew–symmetric. (Hint: The facts we used to enable us to generate examples of symmetric and skew–symmetric matrices, and which you proved in (a) and (b) above, will be useful here.)

3. Graphs and Airline Flight Routes

A *directed graph* is a figure such as

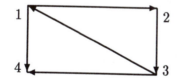

in which a finite number of *vertices* (numbered 1, 2, 3, 4 in this example) are joined by a finite number of *directed edges* (the five arcs with arrows in this example). The *adjacency matrix* (or *vertex matrix*) of a directed graph is defined as follows. Assuming the vertices are numbered 1 to n, the matrix is defined by $A = [a_{ij}]$, where

$$a_{ij} = \begin{cases} 1 & \text{if there is an edge from } i \text{ to } j \\ 0 & \text{otherwise} \end{cases}$$

For the above directed graph,

$$A = \begin{pmatrix} 0 & 1 & 0 & 1 \\ 0 & 0 & 1 & 0 \\ 1 & 0 & 0 & 1 \\ 0 & 0 & 0 & 0 \end{pmatrix}$$

An airline's flight routes can be represented by a directed graph. Each airport is represented by a vertex, and each nonstop flight route is represented by a directed edge from the originating airport to the destination airport. Suppose a small commuter airline covers a portion of California and has the following nonstop flight routes:

Origin	Destination
San Francisco	Fresno
San Francisco	Monterey
Los Angeles	San Francisco
Los Angeles	Sacramento
Sacramento	San Francisco
Sacramento	Fresno
Fresno	Sacramento
Fresno	Los Angeles
Monterey	Los Angeles

(a) Number the cities as follows: (1) Los Angeles, (2) San Francisco, (3) Monterey, (4) Fresno, (5) Sacramento. Draw the directed graph that represents the above set of nonstop flight routes.

(b) Find the adjacency matrix A of this graph, and enter it into MAT-LAB.

(c) Compute A^2. The (i, j) entry of A^2 is the number of routes from airport i to airport j that use exactly two nonstop flight segments. Justify this assertion in detail for the $(2, 1)$ entry of A^2 by explaining what happens when you multiply the second row of A by the first column of A. (Hint: Notice how the 1's match up.)

(d) Compute $A + A^2$. What does the (i, j) entry of this matrix tell you about flight routes?

(e) What is the largest number of flight segments necessary to fly between any two of the five cities? Explain your answer.

(f) The nonstop flight from Sacramento to San Francisco is crucial to this airline. Why? (Hint: Look at what happens in (e) if you remove it.)

(g) Can you add just one nonstop flight to the above schedule so that passengers can fly between any two of the five cities using at most two flight segments? If that is not possible, try adding more flights (but as few as possible) until at most two flight segments are needed. Explain your process.

4. Introduction to Markov Processes

Suppose that a small town has three competing fast food delivery services – Lou's Pizza, Lee's Stir Fry, and Lulu's Tex/Mex. Students in the town make a habit of using one of these food services every Friday evening. Some of them stay with the same service from one week to the next, but some switch around for variety. The directed graph below shows the percentages that stay and the percentages that switch (where P denotes Pizza, S Stir Fry, and T Tex/Mex):

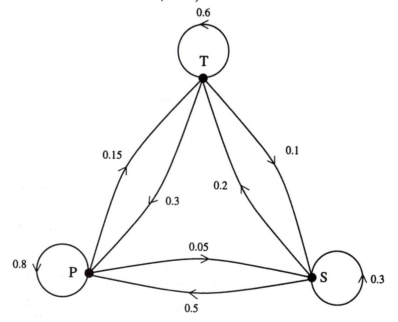

For example, this graph says that, of the students who order Pizza one week, the next week 0.8 of them (80%) again order Pizza, while 0.05 of them (5%) order Stir Fry and 0.15 of them (15%) order Tex/Mex.

(a) If on a particular Friday, 50 students order Pizza, 30 order Stir Fry, and 20 order Tex/Mex, find out how many order each kind of food the next Friday. For example, the number ordering Pizza will be

$$0.8(50) + 0.5(30) + 0.3(20) = 61$$

(b) The above equation together with the two others you found can be written in matrix form $\mathbf{y} = A\mathbf{x}$:

$$\begin{pmatrix} 61 \\ \\ \end{pmatrix} = \begin{pmatrix} 0.8 & 0.5 & 0.3 \\ & & \\ & & \end{pmatrix} \begin{pmatrix} 50 \\ 30 \\ 20 \end{pmatrix}$$

Fill in the eight missing numbers. As a check, note that each column of A should sum to 1.

(c) Describe the meaning of the $(2,3)$ entry of the 3×3 matrix A in (b).

(d) Suppose we change the entries of $\mathbf{x}$ from 50, 30, 20 to any other three initial values and compute the product $A\mathbf{x}$ using this new vector $\mathbf{x}$ but the same matrix A as in (b). Then $A\mathbf{x}$ is again the next week's distribution of customers. Use this idea to explain what $A^2\mathbf{x}$ gives us. Hint: $A^2\mathbf{x} = A(A\mathbf{x})$.

(e) Compute A^2. What do its entries represent? (Just describe the meaning of the $(2,3)$ entry of A^2; the others are similar.)

(f) So if we start with any initial distribution of customers $\mathbf{x_0}$, we can compute the distribution of customers after one week ($\mathbf{x_1}$), after two weeks ($\mathbf{x_2}$), and so on by

$$\mathbf{x_1} = A\mathbf{x_0}, \quad \mathbf{x_2} = A\mathbf{x_1}, \quad \mathbf{x_3} = A\mathbf{x_2}, \ldots \qquad (2.1)$$

Show that $\mathbf{x_k} = A^k\mathbf{x_0}$ for $k = 1, 2, 3, \ldots$.
The sequence of equations (2.1) is called a *Markov process*, and the matrix A is called the *transition matrix* for the process.

(g) Find A^k for larger and larger values of k until increasing k no longer seems to change the numbers in A^k. (This is a way of approximating $\lim_{k\to\infty} A^k$.) Then compute $\lim_{k\to\infty} A^k\mathbf{x_0}$ for our original distribution vector $\mathbf{x_0} = [50 \ 30 \ 20]^T$. What does this tell us about the distribution of customers?

5. A Property of Stochastic Matrices

The matrix A in the preceding project is said to be *stochastic*, which means that its entries are non-negative and each of its columns sums to 1. Stochastic matrices have many useful properties, some of which are investigated in this project and the next.

(a) If A is the matrix in the preceding project and $\mathbf{x} = [50 \ 30 \ 20]^T$, find $\lim_{k \to \infty} A^k \mathbf{x}$ by computing $A^k \mathbf{x}$ for sufficiently large k. Also, solve the equation $A\mathbf{z} = \mathbf{z}$. (Hint: Write the equation $A\mathbf{z} = \mathbf{z}$ in the form $B\mathbf{z} = 0$ and use the ATLAST command nulbasis(B).) What relationship do you see between these two answers? (If you do not see one, go on to part (b), which should help.)

(b) Find $\lim_{k \to \infty} A^k \mathbf{x}$ for the same matrix A but several different randomly chosen non-negative vectors $\mathbf{x}$. What relationship do you see among all these limits? How are these limits related to a solution $\mathbf{z}$ of $A\mathbf{z} = \mathbf{z}$? To help you see a pattern, choose your vectors $\mathbf{x}$ so their entries add to 1:

    x = rand(3,1)
    x = x/sum(x)

Scale $\mathbf{z}$ in the same way: z = z/sum(z).

(c) Repeat the above experiment for other stochastic matrices A. Does the pattern persist? Use the ATLAST command randstoc(n) to construct random stochastic matrices.

(d) Explain why this relationship holds. Hint: If $\mathbf{y} = \lim_{k \to \infty} A^k \mathbf{x}$, what does $A\mathbf{y}$ equal?

6. More Properties of Stochastic Matrices

(a) If A and B are stochastic matrices, which of the following matrices must also be stochastic: $A + B, A - B, AB, A^{-1}, A^2$? Find out by testing these conclusions on several random stochastic matrices; use the ATLAST command randstoc(n).

(b) If A is a stochastic matrix and $\mathbf{x}$ and $\mathbf{y}$ are vectors such that $\mathbf{y} = A\mathbf{x}$, prove that the sum of the entries of $\mathbf{x}$ equals the sum of the entries of $\mathbf{y}$; that is,

$$\sum_{i=1}^{n} \mathbf{x}_i = \sum_{i=1}^{n} \mathbf{y}_i$$

(c) At least one of the five expressions in (a) is always stochastic. Pick one such, and prove that it is stochastic. Hint: Use the result in (b), and use the fact that the product $A\mathbf{e}_j$ equals the jth column of A (where $\mathbf{e}_j$ denotes the jth column of the identity matrix).

7. A Risky Markov Process

Jack is playing pool with Jim for $1 a game. He has only $2 and decides to play until he goes broke or has $5, at which point he will quit and go out for a pizza with Jim (Dutch treat). Jack knows from past experience that he beats Jim 60% of the time. Will Jack end up with $5 to spend on pizza or will he go broke and have to borrow $5 from Jim? That and other questions will be explored in this project.

As in the project "Introduction to Markov Processes," we can represent the probabilities involved by a directed graph:

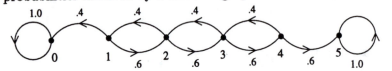

For example, this says that if Jack has $2 before a game, then after the game he will have $1 with probability 0.4 and $3 with probability 0.6. The two loops at the ends are included for completeness; they say that if Jack has $0 or $5, he will continue to have that amount with probability 1. The transition matrix for this Markov process is

$$A = \begin{pmatrix} 1 & 0.4 & 0 & 0 & 0 & 0 \\ 0 & 0 & 0.4 & 0 & 0 & 0 \\ 0 & 0.6 & 0 & 0.4 & 0 & 0 \\ 0 & 0 & 0.6 & 0 & 0.4 & 0 \\ 0 & 0 & 0 & 0.6 & 0 & 0 \\ 0 & 0 & 0 & 0 & 0.6 & 1 \end{pmatrix}$$

(a) Let $\mathbf{x}_0 = [0\ 0\ 1\ 0\ 0\ 0]^{\mathrm{T}}$, which we interpret as saying that initially Jack has $2 with probability 1. (The six entries of $\mathbf{x}_0$ represent the probability that Jack has $0, $1, $2, $3, $4, and $5, respectively.)

Compute $x_1 = Ax_0$. What do the entries of x_1 represent? Compute $x_2 = Ax_1$. What do the entries of x_2 represent?

(b) Find $\lim_{k \to \infty} A^k x_0$ by computing $A^k x_0$ for large k. What do the entries of this limit represent? Will Jack have $5 to spend on pizza? Explain your answer.

(c) If Jack started with a different amount of money, such as $1, $3, or $4, what would his chances be for winning pizza money?

(d) Solve $Az = z$ by using the ATLAST command **nulbasis**. What relationship do you see between the solutions of this equation and the limits in (b) and (c)? How does this pattern compare with the pattern you observed in the project "Introduction to Markov Processes"?

8. The Leslie Population Model

Suppose we are studying an animal population whose population size is determined entirely by the number of females of each age group. Suppose that each one-year-old female produces, on average, three female offspring, and each two-year-old female produces one female offspring. Suppose further that, on average, 37% of the newborn females live to be one year old, 30% of the one-year-olds live to be two years old, and none of the two-year-olds live to be three.

Then we can compute the number of females in each age group from the previous year's data:

$$\begin{aligned}
(\text{current age } 0) &= 3 \times (\text{previous age } 1) + 1 \times (\text{previous age } 2) \\
(\text{current age } 1) &= .37 \times (\text{previous age } 0) \\
(\text{current age } 2) &= .30 \times (\text{previous age } 1)
\end{aligned}$$

In matrix notation, this can be written

$$y_k = Ay_{k-1} \tag{2.2}$$

where

$$A = \begin{pmatrix} 0 & 3 & 1 \\ .37 & 0 & 0 \\ 0 & .30 & 0 \end{pmatrix} \tag{2.3}$$

and

$$\mathbf{y}_k = \begin{pmatrix} \text{number of females of age 0 in year } k \\ \text{number of females of age 1 in year } k \\ \text{number of females of age 2 in year } k \end{pmatrix}$$

(a) Suppose that initially there are 1 million females in each age group. We will write this as $\mathbf{y}_0 = [1\ 1\ 1]^T$, where 1 means 1 million. Compute $\mathbf{y}_1, \mathbf{y}_2, \mathbf{y}_3, \mathbf{y}_4, \mathbf{y}_5$ using formula (2.2).

(b) Explain why $\mathbf{y}_k = A^k \mathbf{y}_0$.

(c) Plot the population in each of the three age groups by using the following MATLAB commands (including the quote (') in the **plot** command):

```
M = [y1, y2, y3, y4, y5]
plot(M′)
```

Note that the populations of the three age groups all move up and down but the trend is generally upward.

(d) Plot the same three population curves for 100 years rather than 5 years, and describe how the populations change. Here is MATLAB code for creating the plot quickly:

```
M = [ ], y = ones(3,1)
for i = 1:100, y = A*y; M = [M,y]; end
plot(M′)
```

A matrix of the form

$$\begin{pmatrix} a_1 & a_2 & a_3 & \cdots & a_{n-1} & a_n \\ p_1 & 0 & 0 & \cdots & 0 & 0 \\ 0 & p_2 & 0 & \cdots & 0 & 0 \\ & & \ddots & & & \\ & & & \ddots & & \\ 0 & 0 & 0 & \cdots & p_{n-1} & 0 \end{pmatrix}$$

is called a *Leslie matrix* and the corresponding population model is a *Leslie model*. (The matrix (2.3) above is a Leslie matrix.) The numbers a_i are the average number of female offspring for each female in that age group, and the numbers p_i are the percent in that age group that survive to the next age group.

(e) Find the Leslie matrix for the following animal population problem, and carry out an investigation of the Leslie model, as you did in (c) and (d).

On average, each one-year-old female produces two female offspring, and each two-year-old female produces two more female offspring. On average, 5/13 of the newborn females live to be one year old, 3/10 of the one-year-olds live to be two years old, 1/4 of the two-year-olds live to be three, and none live to be four.

9. Electrical Networks

A simple electrical circuit consists of electrical sources (such as batteries), resistors (such as light bulbs), and paths (such as wires) over which electrical current flows. Three measurable quantities associated with such a circuit are *electrical potential* (often measured in volts), *resistance* (often measured in ohms), and *current* (often measured in amperes). An electrical source creates a voltage increase and causes current to flow; a resistor causes voltage to drop. Here is a picture of a particular circuit that happens to be called a *Wheatstone bridge.*

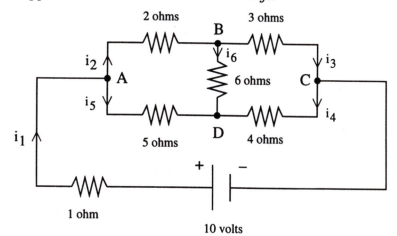

Wheatstone bridge circuit

The symbols have the following meanings:

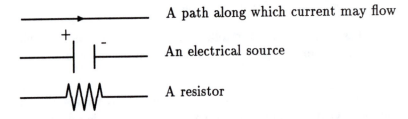

A path along which current may flow

An electrical source

A resistor

We will find the six unknown currents: $i_1, i_2, i_3, i_4, i_5, i_6$. (We pick the directions of flow arbitrarily; some of the currents may therefore turn out to be negative.) To find the currents, we need three physical laws about electrical circuits:

Ohm's Law: The voltage drop V across a resistor is the product of the current i passing through it and its resistance R; that is, $V = iR$.

Kirchoff's Current Law: At every node (i.e., meeting of paths), the sum of the incoming currents equals the sum of the outgoing currents.

Kirchoff's Voltage Law: Around every closed loop, the sum of the voltage increases equals the sum of the voltage drops.

Applying Kirchoff's Current Law to the above example, we get four equations corresponding to the four nodes A, B, C, D:

$$i_1 = i_2 + i_5$$
$$i_2 = i_3 + i_6$$
$$i_3 = i_4 + i_1$$
$$i_4 + i_5 + i_6 = 0 \qquad (2.4)$$

(Note, however, that the fourth equation is redundant; it is the sum of the first three equations.)

To apply Kirchoff's Voltage Law, we denote the voltage drops across the six resistors by $V_1, \ldots, V_6$. Applying the law to the three small loops ABD, BCD, and CAD, we get

$$V_2 + V_6 - V_5 = 0$$
$$V_3 + V_4 - V_6 = 0$$
$$V_1 + V_5 - V_4 = 10 \qquad (2.5)$$

Using Ohm's Law, these equations become

$$2i_2 + 6i_6 - 5i_5 = 0$$
$$3i_3 + 4i_4 - 6i_6 = 0$$
$$i_1 + 5i_5 - 4i_4 = 10 \qquad (2.6)$$

(Other loops yield additional equations, but these are again redundant.)

(a) Solve the system of seven equations (2.4) and (2.6) for the six unknowns $i_1, \ldots, i_6$ using MATLAB's backslash operator \.

(b) Use the above method to find the currents $i_1, i_2, i_3, i_4, i_5, i_6$ in the circuit

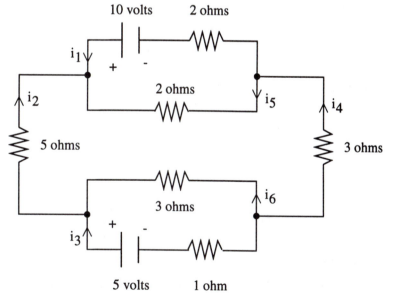

The method we used above has one significant disadvantage. If we change a single resistance anywhere in the network, the whole calculation must be redone. The next method separates the network's "topology" (i.e., its paths and connections) from the numerical values of the resistances. We demonstrate the method on our Wheatstone bridge example.

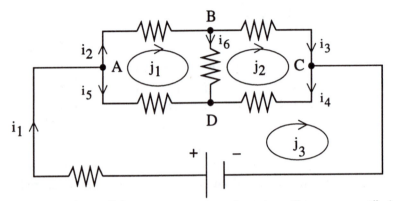

First, in each small loop, we suppose there is a "loop current"; in our example, these are j_1 (in loop ABD), j_2 (in loop BCD), and j_3 (in loop CAD). By expressing each of the currents $i_1, \ldots, i_6$ in terms of the loop currents, we get a system of linear equations:

$$
\begin{aligned}
i_1 &= j_3 \\
i_2 &= j_1 \\
i_3 &= j_2 \\
i_4 &= j_2 - j_3 \\
i_5 &= j_3 - j_1 \\
i_6 &= j_1 - j_2
\end{aligned}
$$

In matrix form, this is

$$
\begin{pmatrix} i_1 \\ i_2 \\ i_3 \\ i_4 \\ i_5 \\ i_6 \end{pmatrix} = \begin{pmatrix} 0 & 0 & 1 \\ 1 & 0 & 0 \\ 0 & 1 & 0 \\ 0 & 1 & -1 \\ -1 & 0 & 1 \\ 1 & -1 & 0 \end{pmatrix} \begin{pmatrix} j_1 \\ j_2 \\ j_3 \end{pmatrix}
$$

or, more succinctly, $I = MJ$. (Note that the columns of M correspond to the topology of the small loops. For example, when we go around the loop ABD, we go in the same direction as currents i_2 and i_6 and in the opposite direction as i_5. This corresponds to the positions of the 1's and -1's in the first column of M.)

The equation $I = MJ$ is essentially Kirchoff's Current Law. Recall that Kirchoff's Voltage Law resulted in equation (2.5), which we can also write in matrix form:

$$\begin{pmatrix} 0 \\ 0 \\ 10 \end{pmatrix} = \begin{pmatrix} 0 & 1 & 0 & 0 & -1 & 1 \\ 0 & 0 & 1 & 1 & 0 & -1 \\ 1 & 0 & 0 & -1 & 1 & 0 \end{pmatrix} \begin{pmatrix} V_1 \\ V_2 \\ V_3 \\ V_4 \\ V_5 \\ V_6 \end{pmatrix}$$

or, more succinctly, $E = M^T V$. It is no accident that the coefficient matrix here is exactly the transpose of the coefficient matrix M in the earlier equation $I = MJ$.

Finally, we use Ohm's Law to relate V and I:

$$V = \begin{pmatrix} V_1 \\ V_2 \\ V_3 \\ V_4 \\ V_5 \\ V_6 \end{pmatrix} = \begin{pmatrix} 1i_1 \\ 2i_2 \\ 3i_3 \\ 4i_4 \\ 5i_5 \\ 6i_6 \end{pmatrix} = \begin{pmatrix} 1 & 0 & 0 & 0 & 0 & 0 \\ 0 & 2 & 0 & 0 & 0 & 0 \\ 0 & 0 & 3 & 0 & 0 & 0 \\ 0 & 0 & 0 & 4 & 0 & 0 \\ 0 & 0 & 0 & 0 & 5 & 0 \\ 0 & 0 & 0 & 0 & 0 & 6 \end{pmatrix} \begin{pmatrix} i_1 \\ i_2 \\ i_3 \\ i_4 \\ i_5 \\ i_6 \end{pmatrix}$$

or, more succinctly, $V = RI$, where R is the diagonal matrix of resistances.

(c) From the three equations above, namely,

$$I = MJ, \quad E = M^T V, \quad V = RI$$

derive

$$E = M^T R M J \tag{2.7}$$

(d) For the Wheatstone bridge example, use the above method to find the loop currents J and then the currents I. Specifically, (i) construct E, M, and R; (ii) compute the coefficient matrix $M^T R M$; (iii) solve equation (2.7) using MATLAB's backslash operator \; and (iv) compute I from $I = MJ$.

(e) Use our second method to find the currents in question (b).

(f) [Optional; for students who have studied the project "Graphs and Incidence Matrices" in Chapter 3.] Construct the incidence matrix for the directed graph underlying the drawing of the Wheatstone bridge. Show that the columns of the matrix M above form a basis of the null space of the incidence matrix.

10. Secret Codes and Matrix Multiplication

One way to disguise a message is to replace each letter in the alphabet by
some other letter and to use this substitution throughout this message.
For example, if we replace each letter by the next one in alphabetical
order and replace Z by A, the message

    THIS IS A SECRET MESSAGE

becomes

    UIJT JT B TFDSFU NFTTBHF

However, substitution codes such as this are easy to break because of
the differing frequencies with which letters appear in English sentences.
We can disguise a message more effectively by converting each letter to
a number and then multiplying the numbers by a matrix to scramble the
letters more thoroughly. In this project, we will use Table 2.1 to convert
the 26 letters of the alphabet and the space, period, and question mark
into the 29 numbers 0, 1, 2, ..., 28 (29 is a prime number).

A	B	C	D	E	F	G	H	I	J
0	1	2	3	4	5	6	7	8	9
K	L	M	N	O	P	Q	R	S	T
10	11	12	13	14	15	16	17	18	19
U	V	W	X	Y	Z		.	?	
20	21	22	23	24	25	26	27	28	

Table 2.1: Numerical conversion

Example: Suppose our *enciphering* matrix (the matrix we use to disguise
our message) is

$$A = \begin{pmatrix} 1 & 2 \\ 3 & 5 \end{pmatrix}$$

Our enciphering scheme is the following:

(i) Use Table 2.1 to convert the message to numbers.

(ii) Group the numbers by twos and multiply each pair of numbers by
   the matrix A.

Let's employ this scheme to encipher the message "THIS IS A SECRET
MESSAGE."

The matrix M below is the result of using Table 2.1 to convert this message to numbers, and C is the enciphered message. The **rem** command replaces each number in C by its remainder after division by 29. Thus the enciphered message contains only the numbers $0, 1, 2, \cdots, 28$, which can then be converted back to letters using Table 2.1.

```
A = [1 2;3 5];
M = [19 7 8 18 26 8 18 26 0 26 18 4 2 17 4 19 26 12 4 18 18 0 6 4];
M = reshape(M,2,length(M)/2)
    M =
        19   8 26 18   0 18   2   4 26   4 18   6
         7 18   8 26 26   4 17 19 12 18   0   4
C = A*M
    C =
        33   44   42   70   52 26 36   42   50   40 18 14
        92 114 118 184 130 74 91 107 138 102 54 38
C = rem(C,29)
    C =
        4 15 13 12 23 26   7 13 21 11 18 14
        5 27   2 10 14 16   4 20 22 15 25   9
```

How can we decipher C? By applying A^{-1} to it, of course! This has to be done somewhat carefully, because we want to use only the numbers $0, 1, 2, \cdots, 28$. Thus, since A^{-1} has some entries not in this range

$$A^{-1} = \begin{pmatrix} -5 & 2 \\ 3 & -1 \end{pmatrix}$$

we will use in its place a matrix B whose entries are all in this range and which have the same remainders after division by 29

$$B = \begin{pmatrix} 24 & 2 \\ 3 & 28 \end{pmatrix}$$

B will be our *deciphering* matrix.

(a) Use the matrix B to convert the enciphered message C back to the original message M.

(b) Make up your own 2×2 enciphering matrix A, using only the numbers $0, 1, 2, \cdots, 28$, and so that A^{-1} has only integer entries. (Hint: Choose A so its determinant is ± 1.) Then change A^{-1} to B, as we did above, so its entries use only $0, 1, 2, \cdots, 28$. Compute AB and BA. How are these products related to the identity matrix?

(c) (This part requires partners.) Make up a message M and encipher it as we did in the example above but using your own enciphering matrix A. Give your partner the enciphered message C, and challenge him or her to decipher it. Do not show your partner your matrices A and B; however, as a clue, provide the first four letters of your original message. (If the determinant of the 2×2 matrix corresponding to these four letters is a multiple of 29, this clue will not be sufficient; change your message.)

(d) Explain the mathematics you used to decipher your partner's message.

Note: There is a mathematical theory governing the algebra of numbers formed by working with remainders after division by a positive integer (such as 29). It is called *modular arithmetic* and can be found in textbooks on number theory or discrete mathematics. If you are familiar with this theory, you might think about the following extension of our methods. Let p be any prime and n be any integer greater than 1. Construct an $n \times n$ enciphering matrix A whose entries use only the numbers $0, 1, \cdots, p-1$ and whose determinant is not equal to 0 modulo p. How would you find a deciphering matrix B which satisfies these same conditions and which when multiplied by A equals I modulo p?

11. Permutation Matrices

A *permutation* matrix is a square matrix that has exactly one 1 in every row and column and 0's elsewhere. The purpose of this project is to discover many of the properties of permutation matrices.

One way to construct permutation matrices is to permute the rows (or columns) of the identity matrix. For example, we can construct

$$E = \begin{pmatrix} 0 & 0 & 1 & 0 & 0 \\ 1 & 0 & 0 & 0 & 0 \\ 0 & 1 & 0 & 0 & 0 \\ 0 & 0 & 0 & 0 & 1 \\ 0 & 0 & 0 & 1 & 0 \end{pmatrix}$$

by using the MATLAB commands

```
p = [3 1 2 5 4];
E = eye(length(p));
E = E(p,:)
```

The second command creates the 5×5 identity matrix, and the third command uses the vector **p** to permute its rows, so row 3 becomes row 1, row 1 becomes row 2, and so on. We will refer to the vector **p** that gives rise to E as the "associated permutation."

(a) Generate several permutation matrices, including one derived from the permutation $\mathbf{p} = [2\ 1\ 5\ 7\ 3\ 6\ 4]$.

(b) We can compute the permutation vector corresponding to a given permutation matrix by multiplying the matrix on the right by the column vector $[1\ 2\ \dots\ n]^\mathrm{T}$:

```
p=(E*[1:size(E)]')')'
```

Compute the permutation vectors from several permutation matrices, including the 5×5 example we started with.

(c) We now examine the effect of multiplication by permutation matrices.

i. Generate several 5×5 permutation matrices E, and generate several other 5×5 matrices A by using, for example, A = randint(5). Compute the products EA and AE.

ii. Describe the effect on A of multiplication by a permutation matrix E.

iii. For the permutation matrix $E_{\mathbf{p}}$ whose associated permutation vector is $\mathbf{p} = [5\ 2\ 1\ 4\ 3]$ and

$$A = \begin{pmatrix} 4 & -2 & 0 & 1 & 3 \\ 7 & 1 & 1 & -2 & 6 \\ 0 & 4 & 5 & 6 & -5 \\ 1 & 1 & 0 & 1 & 2 \\ 2 & 1 & 9 & 7 & 0 \end{pmatrix}$$

find the products $E_{\mathbf{p}}A$ and $AE_{\mathbf{p}}$. Use your description in ii to find these products; do not use MATLAB and do not actually multiply the matrices.

iv. Prove that your answer to ii is correct.

(d) We now examine the product of two permutation matrices.

i. Generate two 5×5 permutation matrices $E_\mathbf{p}$ and $E_\mathbf{q}$ with associated permutation vectors $\mathbf{p}$ and $\mathbf{q}$. Compute the product $E_\mathbf{p}E_\mathbf{q}$, and observe that it is also a permutation matrix.
ii. Find the permutation $\mathbf{r}$ associated with the product $E_\mathbf{p}E_\mathbf{q}$.
iii. Describe the relation between the permutations $\mathbf{p}$, $\mathbf{q}$, and $\mathbf{r}$.
iv. Test your answer to iii on other products of permutation matrices.
v. Prove that your answer to iii is correct. (Hint: Use your answer to (c)ii.)

(e) We now examine the inverse and transpose of a permutation matrix.

i. Generate a permutation matrix $E_\mathbf{p}$ with associated permutation vector $\mathbf{p}$. Compute $E_\mathbf{p}^{-1}$ and $E_\mathbf{p}^T$ and observe that they are also permutation matrices.
ii. Find the permutations $\mathbf{q}$ and $\mathbf{r}$ associated with $E_\mathbf{p}^{-1}$ and $E_\mathbf{p}^T$, respectively.
iii. Describe the relation between the permutations $\mathbf{p}$, $\mathbf{q}$, and $\mathbf{r}$.
iv. Test your answer to iii on other permutation matrices.
v. Prove that your answer to iii is correct.

12. Elementary Matrices and Row Reduction

An *elementary matrix* is any matrix that one gets by applying a single elementary row operation to the identity matrix. For example, the following three elementary matrices were obtained from the 4×4 identity matrix:

$$\begin{pmatrix} 0 & 0 & 1 & 0 \\ 0 & 1 & 0 & 0 \\ 1 & 0 & 0 & 0 \\ 0 & 0 & 0 & 1 \end{pmatrix}, \begin{pmatrix} 1 & 0 & 0 & 0 \\ 0 & 5 & 0 & 0 \\ 0 & 0 & 1 & 0 \\ 0 & 0 & 0 & 1 \end{pmatrix}, \begin{pmatrix} 1 & 0 & 0 & 0 \\ 0 & 1 & 0 & 0 \\ 0 & 0 & 1 & 0 \\ 7 & 0 & 0 & 1 \end{pmatrix} \qquad (2.8)$$

(The row operations were, respectively: Interchange rows 1 and 3; multiply row 2 by -5; multiply row 1 by 7 and add to row 4.) The purpose of this project is to discover some of the properties of elementary matrices and their connection to row reduction of matrices.

Here are two ways to construct elementary matrices using MATLAB; the first uses basic MATLAB commands while the second uses an ATLAST command.

To swap rows i and j of the $n \times n$ identity matrix:

E = eye(n); F = eye(n);
E(i,:) = F(j,:); E(j,:) = F(i,:)

or

E = rowswap(eye(n),i,j)

To multiply row i of the $n \times n$ identity matrix by the scalar c:

E = eye(n);
E(i,i) = c

or

E = rowscale(eye(n),i,c)

To replace row j of the $n \times n$ identity matrix by the sum of row j and c times row i:

E = eye(n);
E(j,i) = c

or

E = rowcomb(eye(n),i,j,c)

(a) Generate several random 4×3 matrices A by the MATLAB command $A = \text{randint}(4,3)$. Compute the products EA for each of the elementary matrices E listed in (2.8). What general pattern do you see? That is, what effect does multiplying a matrix A on the left by an elementary matrix have on the matrix A?

(b) Enter the matrix

$$A = \begin{pmatrix} 1 & -2 & 3 \\ 2 & -6 & 5 \\ -1 & -4 & 0 \end{pmatrix}$$

and reduce it to an upper triangular matrix B by successive multiplications on the left by elementary matrices $E_1, E_2, \cdots$. (Give names to these elementary matrices so you can use them again later.)

(c) From your work in (b), find a matrix P with the property $PA = B$. Describe, in general, for any two row equivalent matrices A and B, how to use elementary matrices to find a matrix P with the property $PA = B$.

(d) Compute the product P*[A,eye(3)]. How is the result of this multiplication related to the matrices A, B, and P? Explain why this always happens.

(e) Test your conclusion in (d) as follows. Enter the 3 × 4 matrix

$$A = \begin{pmatrix} 2 & 4 & -3 & 2 \\ 3 & 1 & 2 & 2 \\ 5 & 5 & -1 & 4 \end{pmatrix}$$

and compute rref([A,eye(3)]). Describe how to read off from this result a matrix P with the property $PA = B$, where B is the reduced row echelon form of A. Check that $PA = B$ is true.

(f) In this final question, we seek a way of using the matrix P in (e) to determine for which vectors $\mathbf{b}$ the system of equations $A\mathbf{x} = \mathbf{b}$ has a solution. For the matrix A in (e), solve $A\mathbf{x} = \mathbf{b}$ for $\mathbf{b} = \mathbf{b}_1 = [4\ 3\ 7]^T$ and $\mathbf{b} = \mathbf{b}_2 = [4\ 3\ 6]^T$ using MATLAB's backslash operator \. Which of the two systems has a solution, and which does not? Now compute the products $P\mathbf{b}_1$ and $P\mathbf{b}_2$. What connection do you see between these products and your conclusion about which system has a solution? (If you do not see a connection, try other vectors $\mathbf{b}$, such as $[1\ 1\ 2]^T$, $[1\ 3\ 2]^T$, $[-2\ 3\ 1]^T$, $[2\ 3\ 1]^T$.) Now describe this phenomenon in general. That is, suppose we are given any system of equations $A\mathbf{x} = \mathbf{b}$ and we are given the matrix P such that PA is the reduced row echelon form of A. By computing $P\mathbf{b}$, how can we tell whether $A\mathbf{x} = \mathbf{b}$ has a solution?

13. Elementary Matrices and Their Inverses

See the description of elementary matrices in the preceding project.

(a) Generate several random 4×3 matrices A by the ATLAST command $A = \mathsf{randint}(4,3,5)$. Compute the products EA for each of the elementary matrices E listed in (2.8). What general pattern do you see? That is, what affect does multiplying a matrix A on the left by an elementary matrix have on the matrix A?

(b) Enter the matrix

$$A = \begin{pmatrix} 1 & -2 & 3 \\ 2 & -6 & 5 \\ -1 & -4 & 0 \end{pmatrix}$$

and reduce it to an upper triangular matrix B by successive multiplications on the left by elementary matrices $E_1, E_2, \cdots$. (Give names to these elementary matrices so you can use them again later.)

(c) From your work in (b), find a matrix P with the property $PA = B$. Describe, in general, for any two row equivalent matrices A and B, how to use elementary matrices to find a matrix P with the property $PA = B$.

(d) Use MATLAB's inv command to compute the inverse of each of the elementary matrices listed in (2.8). Describe how to quickly find the inverse of each of the three types of elementary matrices. In particular, note that the inverse of an elementary matrix is again an elementary matrix; describe the row operation that yields the inverse.

(e) If Q is a nonsingular matrix, explain why Q is row equivalent to the identity matrix. Then, with the help of your conclusion in (d), explain why Q is a product of elementary matrices.

(f) Use your conclusions in (c), (d), and (e) to prove the following theorem: Two $m \times n$ matrices A and B are row equivalent if and only if there is a nonsingular matrix P such that $PA = B$.

(g) Again reduce the matrix A in (b) to an upper triangular matrix B; be careful to use only elementary matrices of the third type (where row j is replaced by the sum of row j and c times row i). As in (c), find a matrix P such that $PA = B$. Note that your elementary matrices are lower triangular and have 1's along the diagonal, which is always true for elementary matrices of the third type. Does it follow that any matrix P that is a product of such elementary matrices will also be lower triangular with 1's along the diagonal? Explain. Is the same true for the inverse of such a matrix? Now state and prove a theorem: Under certain conditions (you must state what these are), a square matrix A can be written as a product $A = LU$, where U is upper

triangular, L is lower triangular, and L has 1's along the diagonal.
(The product LU is called the LU *factorization* of the matrix A.)

14. Partitioned Matrices

MATLAB has powerful commands for selecting submatrices from a matrix and, in the other direction, for constructing a matrix from submatrices. The first few questions in this project are intended simply to give you practice using these commands. The later questions explore some mathematical relationships associated with these commands. As we will see, partitioning a matrix into submatrices can be useful.

(a) Use the ATLAST command A = randint(5,5,9) to create a random 5×5 matrix A with integer entries. Execute each of the following commands and describe in your own words what they do:

    A1 = A(1:3,1:2)
    A2 = A(:,1:2)
    A3 = A(2:4,:)
    A4 = A(2,3:5)

(b) Use a single MATLAB command to create the submatrix of A consisting of rows 2 through 5 and columns 1 through 4.

(c) Form the following matrices:

    B11 = 7*ones(3,1)
    B12 = randint(3,2,5)
    B21 = randint(2,1,5)
    B22 = 6*eye(2)
    B = [B11, B12; B21, B22]

(Notice the similarity with the familiar way in which matrices are often constructed in MATLAB, as in [2, 5; 3, 4].) Print the matrix B. Then draw one horizontal line and one vertical line through B that demonstrate how B is made up of the four matrices B11, B12, B21, B22. We say that B is a *partitioned matrix*.

(d) Using the method of (c), construct the matrix

$$\begin{pmatrix} 3 & 0 & 0 & 5 & 5 \\ 0 & 3 & 0 & 5 & 5 \\ 0 & 0 & 3 & 5 & 5 \\ 4 & 4 & 4 & 0 & 0 \\ 4 & 4 & 4 & 0 & 0 \end{pmatrix}$$

(e) From the matrix A in (a), construct the following submatrices:

A11 = A(1:2,1:3)
A12 = A(1:2,4:5)
A21 = A(3:5,1:3)
A22 = A(3:5,4:5)

Then compute

[A11*B11+A12*B21,A11*B12+A12*B22;
        A21*B11+A22*B21,A21*B12+A22*B22]
A*B

Describe what you have observed as a result of these commands; formulate your observation as a clear statement of how to multiply partitioned matrices.

(f) Use the command C = randint(4,4,2) to construct a random 4 × 4
matrix C with integer entries, and partition C as follows:

C11 = C(1:2,1:2)
C12 = C(1:2,3:4)
C21 = C(3:4,1:2)
C22 = C(3:4,3:4)

(If C11 turns out to be singular, do these steps over.) Using the
method of (c), form the matrices

$$L = \begin{pmatrix} I & O \\ C_{21}C_{11}^{-1} & I \end{pmatrix}, \quad U = \begin{pmatrix} C_{11} & C_{12} \\ O & C_{22} - C_{21}C_{11}^{-1}C_{12} \end{pmatrix}$$

where I and O are 2×2. Compute the product LU, and describe what
you observe.

(g) Relate your observation in (f) to part (g) of Project 13.

(h) State your observation in (f) as a theorem, and prove it by using your statement in (e) to multiply L and U.

15. How Many Singular Matrices are There?

The purpose of this project is to discover how common or uncommon it is for a square matrix to be singular, at least for some special types of matrices. The matrices we use will be generated by the ATLAST command A = randint(n,n,k), where k and n are positive integers. This command produces random $n \times n$ matrices with integer entries in the interval $[-k, k]$. We will only consider 2×2 and 3×3 matrices of this type.

(a) Use MATLAB to generate 100 2×2 matrices of this type, for $k = 1, 2, \cdots, 20$. For each value of k, find the percent that are singular, and enter it in the table below. Here is MATLAB code you may use to generate the matrices and compute the percents:

```
percent = zeros(1,20)
for k=1:20
for i=1:100
    if det(randint(2,2,k)) == 0
        percent(k) = percent(k)+1;
    end
end
end
percent
```

k	1	2	3	4	5	6	7	8	9	10
% singular										
k	11	12	13	14	15	16	17	18	19	20
% singular										

(b) What do these table values indicate about the percent of singular matrices as k increases?

(c) We can picture a column of one of our 2×2 matrices as a vector in the plane with its tail at $(0,0)$ and its head at a point (x, y) that is in the first quadrant and has integer coordinates. (A point with integer coordinates is called a *lattice point*.) When the matrix is singular, how will its two column vectors be related geometrically? Use this geometric description to give a plausible argument (not a proof) to explain the pattern you described in (b).

(d) Repeat the experiment in (a) for 3×3 matrices, and enter the percents in the table below.

k	1	2	3	4	5	6	7	8	9	10
% singular										
k	11	12	13	14	15	16	17	18	19	20
% singular										

(e) What do these table values indicate about the percent of singular matrices as k increases? How does the pattern compare with the pattern for 2×2 matrices?

(f) Give an explanation for this pattern that is similar to the explanation you gave in (c) but uses lattice points in 3-space. How do you visualize the columns of a matrix? How will the columns be related geometrically when the matrix is singular? What does this say about the likelihood that a matrix will be singular?

16. The Geometric Series and Matrix Inverses

The familiar geometric series

$$\frac{1}{1-x} = 1 + x + x^2 + x^3 + \cdots$$

is valid for all real numbers x that satisfy $|x| < 1$. The purpose of this project is to discover conditions under which the analogous formula is valid for square matrices A:

$$(I - A)^{-1} = I + A + A^2 + A^3 + \cdots \tag{2.9}$$

(a) Generate some matrices A by using the MATLAB command A = k*(rand(m)−.5), where m takes on the values 2, 3, 4, and k takes on the values 1/2, 1, 2. For each matrix A, check whether equation (2.9) holds by computing the inverse of $I - A$ and computing some partial sums of the infinite series in equation (2.9). Here is MATLAB code for computing the nth partial sum:

```
S = eye(m); for i = 1:n, S = S+A^i; end, S
```

For what values of m and k does equation (2.9) seem to hold?

(b) How did you decide whether equation (2.9) holds or not? (In effect, this question asks you to come up with a definition of the sum of an infinite series of matrices.)

(c) Consider the special case when $A = kI$, where k is any scalar. Substitute kI for A in equation (2.9) and determine, by a hand computation, the values of k for which equation (2.9) holds.

(d) Consider the special case when $A = kJ$, where J is the *ones* matrix and k is any scalar. Again, determine the values of k for which equation (2.9) holds. (Hint: J^2 is just a scalar times J; find that scalar; then determine J^i for $i > 2$.)

(e) One way to test the validity of equation (2.9) is to check whether

$$I = (I - A)(I + A + A^2 + A^3 + \cdots)$$

Since the distributive law might not hold for infinite series, we check instead whether

$$I = \lim_{n \to \infty} (I - A)(I + A + A^2 + A^3 + \cdots + A^n)$$

Show that this equation is equivalent to

$$\lim_{n \to \infty} A^n = 0 \qquad (2.10)$$

(f) Find a condition on A which guarantees that equation (2.10) holds (and hence that (2.9) holds). Look back at your work in (a), (c), and (d) for clues. (A condition that is not just sufficient but also necessary is much harder to find; it requires the use of eigenvalues.)

Chapter 3

Determinants

3.1 Exercises on Determinants

Determinants and Row Operations

In this group of exercises we investigate the effects of each of the three row operations on the value of the determinant of a matrix.

1. **Row Operation I**

   The first of the three row operations interchanges the order of two rows of a matrix. One can interchange rows i and j of a matrix A by using the ATLAST command **rowswap(A, i, j)**. To see the effects of this row operation set

   ```
   n = 5; i = 2; j = 4;
   A = randint(n)
   B = rowswap(A, i, j)
   d1 = det(A)
   d2 = det(B)
   ```

   How do the determinants **d1** and **d2** compare? Repeat this experiment using different values for n and different values of i and j. Make a conjecture about the effect of row operation I on the value of the determinant of a matrix.

2. **Row Operation II**

   The second row operation multiplies a row of a matrix by a nonzero

scalar c. This operation can be performed using the ATLAST command
rowscale(A, i, c). To see the effects of this operation set

```
n = 4; i = 3; c = 5;
A = randint(n)
B = rowscale(A, i, c)
d1 = det(A)
d2 = det(B)
```

How do d1 and d2 compare? Repeat this experiment using different values
of n, i, and c. Make a conjecture about the effect of row operation II on
the value of the determinant of a matrix.

3. Row Operation III

The third row operation adds a multiple of one row of A to another.
This operation can be performed using the ATLAST command row-
comb(A,i,j,c) which adds c times row i of A to row j of A. To see the
effects of this operation set

```
n = 5; i = 2; j = 3; c = 2;
A = randint(n)
B = rowcomb(A, i, j, c)
d1 = det(A)
d2 = det(B)
```

How do d1 and d2 compare? Repeat this experiment using different values
of n, i, j and c. Make a conjecture about the effect of row operation III
on the value of the determinant of a matrix.

Determinants and Matrix Operations

In these exercises we investigate how determinants relate to matrix operations such as transpose, scalar multiplication, matrix addition and matrix
multiplication.

4. Scalar Multiplication

In this exercise we investigate how the determinants of A and cA compare.

(a) Set

```
n = 2; c = 2;
A = randint(n)
d1 = det(A)
d2 = det(c*A)
```

How do **d1** and **d2** compare? Repeat the experiment with $n = 3, 4, 5$ and also with $c = 3$ and $n = 2, 3, 4, 5$.

(b) Make a conjecture about the relationship between the determinants of A and cA when A is $n \times n$.

(c) Use your conclusion from Exercise 2 to prove your conjecture in part (b).

5. Transpose

For various values of n, generate a random integer $n \times n$ matrix A and compute the determinants of A and A^T. Make a conjecture about how the determinants of A and A^T compare. If you are familiar with mathematical induction, use it to prove your conjecture. Otherwise, answer the following questions which are designed to illustrate the idea behind a mathematical induction proof.

(a) Prove your conjecture in the case $n = 2$.

(b) To see why the conjecture is true for $n = 3$, make up a 3×3 matrix A and compute by hand the cofactor expansion of $\det(A)$ along its first row. Compute also the cofactor expansion of $\det(A^T)$ along its first column. How do the two cofactor expansions compare? If your conjecture is true for $n = 2$, why must it also be true for the case $n = 3$?

(c) Suppose k is an integer for which your conjecture holds (such as 2 or 3) and let B be any $(k + 1) \times (k + 1)$ matrix. How will the cofactor expansion of $\det(B)$ along its first row compare to the cofactor expansion of $\det(B^T)$ along its first column? If your conjecture is true in the case $n = k$, why must it also be true for the case $n = k + 1$?

(d) Using the results from parts (b) and (c), explain why your conjecture must hold in the case $n = 4$. Next explain why it will hold if $n = 5$. Is there any positive integer value of n for which it will not hold? Explain.

6. Matrix Addition and Multiplication

For various values of n, generate random integer $n \times n$ matrices A and B, and in each case compute the following pairs of numbers:

       (i) $\det(A + B)$ and $\det(A) + \det(B)$
      (ii) $\det(AB)$ and $\det(A)\det(B)$

Is it possible to make a conjecture about either $\det(A + B)$ or $\det(AB)$?

7. Inverses

For $n = 2, 3, 4, 5$, generate a random integer matrix A and in each case compute $\det(A)$ and $\det(A^{-1})$. Make a conjecture about the relation between $\det(A)$ and $\det(A^{-1})$. How does this conjecture relate to the one you made in Exercise 6?

Determinants of Special Matrices

This group of exercises involves experimenting with determinants of special matrices.

8. Matrices with Duplicate Rows

In this exercise we consider matrices having two rows that are the same.

(a) Prove that if A is a 2×2 matrix whose rows are identical, then $\det(A) = 0$.

(b) Does the result in part (a) hold for $n \times n$ matrices if $n > 2$? For $n = 3, 4, 5$, construct a matrix with two identical rows by setting

      A = randint(n);
      A(2,:) = A(1,:);

In each case compute the determinant.

(c) If A is a 3×3 matrix with two identical rows, use the cofactor expansion and the result from part (a) to explain why the determinant of A must be 0. If you are familiar with mathematical induction, use it to give a formal proof that if an $n \times n$ matrix A has two rows the same, then $\det(A) = 0$.

(d) Prove that if an A is an $n \times n$ matrix and one row of A is a multiple of another, then $\det(A) = 0$.

9. Skew-Symmetric Matrices

Recall that a matrix A is said to be *skew-symmetric* if $A^{\mathrm{T}} = -A$. We saw in Chapter 2, Section 2 that if B is any square matrix then $B - B^{\mathrm{T}}$ is skew-symmetric. For $n = 2, 3, \ldots, 7$, compute random integer skew-symmetric matrices by setting

```
B = randint(n);
A = B − B′
```

In each case, compute the determinant of the matrix A. There should be an obvious pattern when n is odd. If n is an odd integer, make and prove a conjecture about $n \times n$ skew-symmetric matrices. (Hint: Use your results from Exercises 4 and 5.) Why does your proof fail in the case where n is even?

10. Backwards Identity

Let B_n denote the $n \times n$ *backwards identity* matrix. It can be generated by the command backiden(n). Generate the matrices B_n for $n = 2, 3, 4, \ldots, 9$, and in each case compute its determinant. Make a conjecture about the determinant of B_n. (Consider separately the case where n is of the form $4k$ or $4k + 1$ and the case where n is of the form $4k + 2$ or $4k + 3$).

11. Consecutive Integers Matrices

The $n \times n$ *consecutive integers matrix* can be generated by using the ATLAST command consec(n). The Hankel form of the consecutive integers matrix can be generated by using the ATLAST command hconsec(n).

(a) For various values of n where $n > 2$, generate $n \times n$ consecutive integers matrices in Hankel form. In each case compute the determinant of the matrix. What do you observe? Make a conjecture about consecutive integers matrices in Hankel form.

(b) Generate a 5×5 consecutive integers matrix H in Hankel form. Use the rowcomb command to subtract the second row of H from the third and then use the rowcomb operation again on the resulting matrix to subtract its first row from its second row. Why must the determinant of the resulting matrix be 0? Generalize this example to the $n \times n$ case and use it to prove your conjecture.

(c) Compute the determinants of the ordinary $n \times n$ consecutive integers matrices for various values of n. Does your conjecture from part (a) hold for consecutive integer matrices that are not in Hankel form? If so, prove it.

12. Minimum Matrix

The *minimum matrix* is generated by the ATLAST command minmat(n).

(a) For $n = 2, 3, 4, 5, 6$, generate the $n \times n$ minimum matrix and compute its determinant. Make a conjecture about the determinant of the minimum matrix.

(b) Set M = minmat(4) and use the **rowcomb** operation to reduce M to an upper triangular matrix U. Why must M and U have the same determinant? In general, if the $n \times n$ minimum matrix M is reduced to an upper triangular matrix U using only the **rowcomb** operation, what form will U have?

13. Maximum Matrix

The *maximum matrix* is generated by the ATLAST command maxmat(n).

(a) For $n = 2, 3, 4, 5, 6$, generate the $n \times n$ maximum matrix and compute its determinant. Make a conjecture about the determinant of the maximum matrix.

(b) Set A = maxmat(4) and L = rowcomb(A, 2, 1, −1). Use the **rowcomb** command to subtract the third row of L from its second row. Subtract the fourth row of the resulting matrix from its third row. You should end up with a lower triangular matrix. Why must the determinant of A be equal to the determinant of the lower triangular matrix?

(c) To automate the process in part (b) set

string = 'L = A; for i = 1:n−1,L = rowcomb(L,i+1,i,−1); end, L'

Now to reduce the $n \times n$ maximum matrix to lower triangular form one need only enter the value of n, set A = maxmat(n) and type: eval(string). For $n = 2, 3, 4, 5, 6$ use this method to reduce the maximum matrix to a lower triangular matrix L. Explain your conjecture from part (a) by relating the determinant of the maximum matrix to the determinant of L.

Alphabet Matrices

In this group of exercises we consider determinants of matrices corresponding to letters of the alphabet. The alphabet matrices are composed of 0's and 1's. In each case the 1's are positioned in the form of a letter of the alphabet. For example the 4×4 letter L matrix is given by

$$\begin{pmatrix} 1 & 0 & 0 & 0 \\ 1 & 0 & 0 & 0 \\ 1 & 0 & 0 & 0 \\ 1 & 1 & 1 & 1 \end{pmatrix}$$

14. Letter L,N,X,Z Matrices

Matrices corresponding to the letters L,N,X,Z can all be generated using ATLAST commands. What types of values would you expect to get if you computed the determinants of these alphabet matrices? Choose a few values of n and compute the determinants of some or all of these matrices. What basic property do these four letter matrices have in common?

15. Letter Y Matrix

If n is odd, one can generate an $n \times n$ matrix in the form of the letter Y using the ATLAST command Ymatrix(n).

(a) For $n = 3, 5, 7, 9, 11, 13$ generate the $n \times n$ letter Y matrix and compute its determinant. Make a conjecture about the determinant of the $n \times n$ letter Y matrix for any odd integer n.

(b) Set Y = Ymatrix(5) and use the **rowswap** command twice to transform Y to a lower triangular matrix. How many row operations would it take to transform the 7×7 letter Y matrix to a lower triangular matrix? How is the determinant of the letter Y matrix related to the determinant of the lower triangular matrix?

3.2   Projects on Determinants

Determinants of Special Tridiagonal Matrices

In this group of projects we examine the determinants of some symmetric tridiagonal Toeplitz matrices. A matrix is *tridiagonal* if all of its nonzero entries lie on the main diagonal and the diagonals directly above and below the main diagonal. A *Toeplitz matrix* is a matrix whose entries along any diagonal are constant.

1. Consider $n \times n$ matrices A_n with 2's on the main diagonal and -1's on the diagonals directly above and below the main diagonal. All other entries of A_n are 0. Thus, for example,

$$A_1 = (2), \quad A_2 = \begin{pmatrix} 2 & -1 \\ -1 & 2 \end{pmatrix}, \quad A_3 = \begin{pmatrix} 2 & -1 & 0 \\ -1 & 2 & -1 \\ 0 & -1 & 2 \end{pmatrix}$$

These matrices are easily generated using MATLAB. For example, to generate A_4 set

```
n = 4;
A = toeplitz([2, −1, zeros(1,n − 2)])
```

(a) For $n = 2, 3, 4, 5, 6$, calculate the determinant of A_n. Do you see a pattern emerging? Make a conjecture about the value of $\det(A_n)$.

(b) Show that

$$\det(A_{n+1}) = 2\det(A_n) - \det(A_{n-1}) \qquad (n \geq 3)$$

This type of equation is called a recursion relation.

2. In this project we examine tridiagonal matrices B_n that have 5's on the main diagonal and 2's on the diagonals just above and below the main diagonal. For example,

$$B_4 = \begin{pmatrix} 5 & 2 & 0 & 0 \\ 2 & 5 & 2 & 0 \\ 0 & 2 & 5 & 2 \\ 0 & 0 & 2 & 5 \end{pmatrix}$$

(a) Find a recursion relation for determining $\det(B_{n+1})$. (Try expanding B_{n+1} along its first row or first column.)

(b) For $n = 1, 2, 3, 4, 5$, compute $\det(B_n)$. Verify that

$$\det(B_1) = 1 + 2^2, \quad \det(B_2) = 1 + 2^2 + 2^4, \quad \det(B_3) = 1 + 2^2 + 2^4 + 2^6$$

and that the determinants of B_4 and B_5 follow the same pattern. Give a general formula for $\det(B_n)$. Show how the formula can be derived by using the recursion relation from part (a) and substituting $2^2 + 1$ for 5.

3. Observe that the matrices in Exercises 1 and 2 fit the following pattern: there is a number c on the sub- and super-diagonals, $c^2 + 1$ on the main diagonal, and 0's everywhere else. Denote the $n \times n$ matrix of that form by C_n.

(a) Find a recursion relation for $\det(C_{n+1})$ and make a conjecture about the values of $\det(C_n)$. (This time think about sums of even powers of c.)

(b) Test your conjecture with some further specific values of c. You might want to try some non-integer values such as $c = \frac{3}{4}$ or $c = -\frac{3}{2}$ in order to keep the determinants to reasonable sizes.

(c) Test your conjecture using MATLAB's symbolic toolbox. Enter

$$C = \begin{pmatrix} c^2 + 1 & c & 0 \\ c & c^2 + 1 & c \\ 0 & c & c^2 + 1 \end{pmatrix}$$

in MATLAB as a symbolic matrix and compute its determinant. Does the result agree with your conjecture?

(d) Prove your conjecture from part (a). Do you see any way to generalize your conjecture?

Interpolation

Given a set of points $(x_1, y_1), (x_2, y_2), \ldots, (x_n, y_n)$, we wish to find a polynomial $c(x)$ whose graph passes through all of the points. If such a polynomial $c(x)$ exists, we say that $c(x)$ *interpolates* the given points. In many applications it is desirable to interpolate a set of points with the lowest possible degree polynomial that will do the job.

4. Linear Interpolation

Consider the problem of interpolating two points $(x_1, y_1), (x_2, y_2)$ in the plane. If the two points are distinct they determine a line. This suggests that a linear polynomial is needed. A linear polynomial $c(x) = c_1 x + c_2$ will interpolate the points if and only if

$$c_1 x_1 + c_2 = y_1$$
$$c_1 x_2 + c_2 = y_2$$

This is a linear system of equations in the unknowns c_1 and c_2. The coefficients of the polynomial can be determined by solving the system. The system can be written as a matrix equation

$$\begin{pmatrix} x_1 & 1 \\ x_2 & 1 \end{pmatrix} \begin{pmatrix} c_1 \\ c_2 \end{pmatrix} = \begin{pmatrix} y_1 \\ y_2 \end{pmatrix}$$

It will have a unique solution if and only if its coefficient matrix is nonsingular. Compute by hand the determinant of the coefficient matrix. Under what conditions will there exist a unique polynomial $c(x) = c_1 x + c_2$ that interpolates the points $(x_1, y_1), (x_2, y_2)$?

5. Consider now the problem of interpolating three points

$$(x_1, y_1), (x_2, y_2), (x_3, y_3)$$

If we use a quadratic polynomial $c(x) = c_1 x^2 + c_2 x + c_3$ and require that

$$c(x_i) = y_i \qquad i = 1, 2, 3$$

then as in the previous exercise we can determine the coefficients of $c(x)$ by solving a linear system

$$\begin{pmatrix} x_1^2 & x_1 & 1 \\ x_2^2 & x_2 & 1 \\ x_3^2 & x_3 & 1 \end{pmatrix} \begin{pmatrix} c_1 \\ c_2 \\ c_3 \end{pmatrix} = \begin{pmatrix} y_1 \\ y_2 \\ y_3 \end{pmatrix}$$

Enter the coefficient matrix and right hand side of this system in MAT-LAB as symbolic matrices V and **y** and solve for **c**. Compute the determinant d of the symbolic matrix V. How is the determinant related to the coefficients c_1, c_2 and c_3? To see the determinant in factored form, use the command **factor(d)**. What conditions must the x coordinates of the three points satisfy in order to guarantee that there will be a unique interpolating polynomial of degree 2 or less?

6. Consider now the problem of interpolating four points

$$(x_1, y_1), (x_2, y_2), (x_3, y_3), (x_4, y_4)$$

by a polynomial of degree 3 or less. As in Exercise 5, set up a symbolic linear system to determine the coefficients of the interpolating polynomial. Find the determinant of the coefficient matrix in factored form. What conditions must the x coordinates of the four points satisfy in order to guarantee that there will be a unique interpolating polynomial of degree 3 or less? Make a conjecture about when there will be a unique polynomial of degree n or less that interpolates the points $(x_1, y_1), (x_2, y_2), \ldots, (x_{n+1}, y_{n+1})$.

7. The coefficient matrix for the interpolation problem with $n + 1$ points is

$$V = \begin{pmatrix} x_1^n & x_1^{n-1} & \cdots & x_1 & 1 \\ x_2^n & x_2^{n-1} & \cdots & x_2 & 1 \\ \vdots & & & & \\ x_{n+1}^n & x_{n+1}^{n-1} & \cdots & x_{n+1} & 1 \end{pmatrix}$$

A matrix of this form is called a *Vandermonde matrix*. For a given numeric vector **x**, one can generate the corresponding Vandermonde matrix by using the MATLAB command **vander(x)**. For example, if we set x=[2;3;5;10], the command **vander(x)** will result in the matrix

$$\begin{pmatrix} 8 & 4 & 2 & 1 \\ 27 & 9 & 3 & 1 \\ 125 & 25 & 5 & 1 \\ 1000 & 100 & 10 & 1 \end{pmatrix}$$

Given the points $(-1, 3), (0, 1), (2, 5), (4, 3)$, and $(5, 7)$, use the **vander** command to generate the coefficient matrix of a linear system $V\mathbf{c} = \mathbf{y}$. Use the \ operation to solve the system for **c**. The coordinates of **c** will be the coefficients of the polynomial that interpolates the given points. To graph the points and the interpolating polynomial, use the commands

```
t = [-1:0.1:5];
z = polyval(c,t);
plot(t,z,x,y,'x')
```

Chapter 4

Vector Space Concepts

4.1 Exercises on Vector Space Concepts

Note: Exercises 1-4 do not require use of a computer. In Exercises 1 and 2, picture all vectors $[x \ y \ z]^T$ as directed line segments in 3-space with tail at $(0,0,0)$ and head at (x, y, z).

Geometric Views of Basic Concepts

1. In this exercise, we explore the geometric meaning of linear independence and dependence of vectors in $\mathbf{R}^3$.

   (a) Find a geometric necessary and sufficient condition for two vectors in $\mathbf{R}^3$ to be linearly dependent. Then draw two linearly dependent vectors and two linearly independent vectors.

   (b) Suppose $\mathbf{v}_1$ and $\mathbf{v}_2$ are linearly independent vectors in $\mathbf{R}^3$. Find a geometric necessary and sufficient condition for a third vector $\mathbf{v}_3$ to be a linear combination of $\mathbf{v}_1$ and $\mathbf{v}_2$.

   (c) State a geometric necessary and sufficient condition for three vectors in $\mathbf{R}^3$ to be linearly dependent. (Your answer to (b) will help here.) Draw three linearly dependent vectors and three linearly independent vectors.

2. In this exercise, we investigate subspaces of $\mathbf{R}^3$ geometrically.

   (a) If $\mathbf{v}_1$ is a nonzero vector in $\mathbf{R}^3$, describe the span of $\{\mathbf{v}_1\}$ geometrically.

(b) If $\mathbf{v}_1$ and $\mathbf{v}_2$ are linearly independent vectors in $\mathbf{R}^3$, describe the span of $\{\mathbf{v}_1, \mathbf{v}_2\}$ geometrically. (See parts (a) and (b) of the preceding exercise.)

Definitions of Basic Concepts

3. Consider the matrices

$$A = \begin{pmatrix} 1 & -2 \\ 3 & -6 \end{pmatrix}, \quad B = \begin{pmatrix} 1 & -2 \\ 0 & 0 \end{pmatrix}$$

A single row operation reduces the matrix A to the matrix B. Show that A and B have the same row space but different column spaces.

4. These are questions to be answered without any calculation at all, except perhaps for a little mental arithmetic. Explain, briefly, why each of your answers is correct. All the questions refer to the vectors

$$\mathbf{v}_1 = \begin{pmatrix} 1 \\ 0 \\ -2 \end{pmatrix}, \quad \mathbf{v}_2 = \begin{pmatrix} 3 \\ -1 \\ 0 \end{pmatrix}$$

(a) Show that $\mathbf{v}_1$ is in the null space of

$$A = \begin{pmatrix} 2 & 4 & 1 \\ -4 & 5 & -2 \end{pmatrix}$$

(b) Show that $[0 \ -1 \ 6]^{\mathrm{T}}$ is in the span of $\{\mathbf{v}_1, \mathbf{v}_2\}$.
(c) Find a vector that is not in the span of $\{\mathbf{v}_1, \mathbf{v}_2\}$.

5. (a) Suppose we want to study some given vectors $\mathbf{v}_1, \cdots, \mathbf{v}_k$ in $\mathbf{R}^n$ and that we have created a MATLAB matrix V ($n \times k$) whose columns are $\mathbf{v}_1, \cdots, \mathbf{v}_k$. For each of the following fundamental questions about the vectors $\mathbf{v}_1, \cdots, \mathbf{v}_k$, describe how to use MATLAB to answer the question, and explain why your method works. (Find methods that are as quick as possible and require little or no hand computation.)

    i. Are $\mathbf{v}_1, \cdots, \mathbf{v}_k$ linearly independent?
    ii. Do $\mathbf{v}_1, \cdots, \mathbf{v}_k$ span $\mathbf{R}^n$?

iii. Suppose **w** is another vector in $\mathbf{R}^n$ and that we have it stored in a $n \times 1$ MATLAB matrix. Is **w** in the span of $\{\mathbf{v}_1, \cdots, \mathbf{v}_k\}$? If so, express **w** as a linear combination of $\mathbf{v}_1, \cdots, \mathbf{v}_k$.

iv. Find a basis of the span of $\{\mathbf{v}_1, \cdots, \mathbf{v}_k\}$ and the dimension of this span.

(b) Answer the four questions in (a) for the matrix

$$V = \begin{pmatrix} 2 & 3 & 3 \\ -3 & 4 & 1 \\ 0 & 1 & 3 \\ -1 & 2 & -3 \end{pmatrix}$$

In part iii, use the vectors $\mathbf{w}_1 = [1 \ 0 \ 0 \ 1]^T$, $\mathbf{w}_2 = [1 \ 1 \ 0 \ 1]^T$.

(c) Answer the four questions in (a) for the matrix

$$V = \begin{pmatrix} 4 & 2 & 3 & 2 & 1 \\ 3 & -4 & -3 & 2 & 0 \\ -1 & -4 & 1 & 3 & -1 \\ 3 & -3 & 4 & -2 & 1 \end{pmatrix}$$

In part iii, use the vectors $\mathbf{w}_1$ and $\mathbf{w}_2$ above.

Subspaces Associated with a Matrix

Exercises 6, 7, and 8 all refer to the matrix

$$A = \begin{pmatrix} 1 & 2 & 5 & -2 & 1 \\ -1 & 2 & 3 & 2 & 3 \\ 4 & -1 & 2 & -1 & 2 \\ 2 & -1 & 0 & 3 & 4 \end{pmatrix}$$

6. (a) Reduce A to a row echelon form using step-by-step Gaussian elimination with the aid of the ATLAST commands rowswap, rowscale, and rowcomb. Write down the row operations you performed in the order in which you performed them.

(b) Using the echelon matrix you found in (a), find a basis of the row space of A.

(c) With the aid of your list of row operations in (a), express each of the original rows of A as a linear combination of the basis vectors in (b).

7. (a) Suppose M is an $m \times n$ matrix of rank r. From this information alone, find

    i. The dimension of the row space of M;
    ii. The dimension of the column space of M;
    iii. The dimension of the null space of M;
    iv. The dimension of the null space of M^{T}.

  (b) Use MATLAB to calculate the rank of the matrix A above. From that one calculation, find the four dimensions listed in (a).

8. For the matrix A above, find a basis for each of the following subspaces, using as few MATLAB commands as possible:

  (a) The row space of A;
  (b) The column space of A;
  (c) The null space of A;
  (d) The null space of A^{T}.

Tell what commands you used and how you got your answers from them.

9. Construct several $m \times n$ matrices of rank r by ATLAST commands of the form

$$M = \mathsf{randint(m,n,5,r)}$$

for various values of m, n, and r. For each such matrix M, compute the nullity of M. (The *nullity* of M is the dimension of the null space of M; the ATLAST command **nulbasis(M)** and the MATLAB command **null(M)** both produce a matrix whose columns form a basis of the null space of M.) Describe the pattern you see.

Coordinates with Respect to a Basis

10. In this exercise, we investigate properties of the coordinates of a vector with respect to a basis.

  (a) The vectors

$$\mathbf{u}_1 = \begin{pmatrix} 3 \\ 5 \\ -1 \end{pmatrix}, \ \mathbf{u}_2 = \begin{pmatrix} -2 \\ 0 \\ 3 \end{pmatrix}, \ \mathbf{u}_3 = \begin{pmatrix} 3 \\ 2 \\ -3 \end{pmatrix}$$

form a basis of $\mathbf{R}^3$. Express the vector $\mathbf{w} = [-3 \ 1 \ 5]^{\mathrm{T}}$ as a linear combination of $\mathbf{u}_1$, $\mathbf{u}_2$, $\mathbf{u}_3$:

$$\mathbf{w} = c_1 \mathbf{u}_1 + c_2 \mathbf{u}_2 + c_3 \mathbf{u}_3$$

The coefficients c_1, c_2, c_2 are called the *coordinates* of $\mathbf{w}$ with respect to the basis $\{\mathbf{u}_1, \mathbf{u}_2, \mathbf{u}_3\}$; the vector $[c_1 \ c_2 \ c_3]^{\mathrm{T}}$ is called the *coordinate vector* of $\mathbf{w}$, and is denoted $\mathbf{w}_{\mathbf{u}}$.

(b) Find the coordinate vector of $\mathbf{w}$ with respect to the basis

$$\mathbf{v}_1 = \begin{pmatrix} -3 \\ 1 \\ 5 \end{pmatrix}, \ \mathbf{v}_2 = \begin{pmatrix} 4 \\ -2 \\ -7 \end{pmatrix}, \ \mathbf{v}_3 = \begin{pmatrix} 8 \\ 1 \\ -11 \end{pmatrix}$$

Denote the coordinate vector $\mathbf{w}_{\mathbf{v}}$.

(c) Find the matrix A whose columns are the coordinate vectors of $\mathbf{u}_1$, $\mathbf{u}_2$, $\mathbf{u}_3$ with respect to the basis $\{\mathbf{v}_1, \mathbf{v}_2, \mathbf{v}_3\}$. Find the matrix B whose columns are the coordinate vectors of $\mathbf{v}_1$, $\mathbf{v}_2$, $\mathbf{v}_3$ with respect to the basis $\{\mathbf{u}_1, \mathbf{u}_2, \mathbf{u}_3\}$. (Hint: A and B can be computed directly from the matrices U, V, U^{-1}, V^{-1}, where U is the matrix with columns $\mathbf{u}_1$, $\mathbf{u}_2$, $\mathbf{u}_3$, and V is the matrix with columns $\mathbf{v}_1$, $\mathbf{v}_2$, $\mathbf{v}_3$.)

(d) Compute various products using the matrices A, B, $\mathbf{w}_{\mathbf{u}}$, and $\mathbf{w}_{\mathbf{v}}$. (For example, compute AB and $A\mathbf{w}_{\mathbf{u}}$.) State all the relationships you find among these four matrices. Explain why they work out this way.

Methods for Constructing a Basis

11. In a sense, finding a basis for a vector space V is easy — so easy that you can simply select a set of vectors in V at random (with certain constraints that you will soon discover), and usually you will find that you have selected a basis. In this exercise, you will find bases of $\mathbf{R}^4$ by a random process.

(a) Construct several $4 \times m$ matrices A by using ATLAST commands of the form

$$A = \mathsf{randint}(4, m) \qquad\qquad (4.1)$$

where $m = 2, 3, 4, 5$. Find out whether the columns of A form a basis of $\mathbf{R}^4$ by the following techniques. The columns of A are linearly independent if and only if the null space of A consists of only the zero vector (in which case the null space has no basis); so use the ATLAST command $\mathsf{nulbasis(A)}$ or the MATLAB command $\mathsf{null(A)}$. The columns of A span $\mathbf{R}^4$ if and only if the dimension of the column space is 4 (i.e., $\mathrm{rank}(A) = 4$); so use the command $\mathsf{rank(A)}$.

(b) What pattern do you see? That is, under what constraints do you expect the command (4.1) to produce a basis of $\mathbf{R}^4$? Explain why this constraint is needed.

12. In this exercise you will explore a method for finding a basis of the span of $\{\mathbf{v}_1, \cdots, \mathbf{v}_k\}$ in which the basis vectors are taken from the set $\{\mathbf{v}_1, \cdots, \mathbf{v}_k\}$. Our method will be to discover which of the vectors $\mathbf{v}_i$ can be expressed as linear combinations of the others; then we throw out all such $\mathbf{v}_i$; the remaining vectors form a basis.

We will find a basis for the span of the columns $\{\mathbf{v}_1, \mathbf{v}_2, \mathbf{v}_3, \mathbf{v}_4\}$ of the matrix

$$V = \begin{pmatrix} -4 & -3 & -2 & -5 \\ -3 & -2 & -1 & 6 \\ 1 & 3 & 5 & 7 \\ -4 & 1 & 6 & 3 \end{pmatrix}$$

(a) Find the reduced row echelon form of V, and show that a basis of the null space of V is $[1 \ -2 \ 1 \ 0]^{\mathrm{T}}$. Thus

$$[\mathbf{v}_1 \ \mathbf{v}_2 \ \mathbf{v}_3 \ \mathbf{v}_4] \begin{pmatrix} 1 \\ -2 \\ 1 \\ 0 \end{pmatrix} = 0$$

and so

$$\mathbf{v}_1 - 2\mathbf{v}_2 + \mathbf{v}_3 = 0$$

Since v_3 is a linear combination of v_1 and v_2, we throw out v_3. The remaining vectors $\{v_1, v_2, v_4\}$ form a basis for the span of $\{v_1, v_2, v_3, v_4\}$. (Notice also that the pivots of A are in columns 1, 2, 4; this gives us an even quicker way to spot the basis.)

(b) Use the method in (a) to find a basis for the span of the columns $\{v_1, v_2, v_3, v_4, v_5\}$ of the matrix

$$V = \begin{pmatrix} 3 & 1 & 7 & 5 & 5 \\ 4 & 6 & 0 & -2 & -1 \\ 1 & 1 & 1 & -3 & -2 \\ 1 & -3 & 9 & 1 & 2 \end{pmatrix}$$

Check that you have the right number of basis vectors by finding the rank of V. (Recall that the rank of V equals the dimension of the column space of V.)

13. In this exercise you will explore a method for extending a linearly independent set of vectors $\{v_1, \cdots, v_k\}$ in $\mathbf{R}^n$ to a basis of $\mathbf{R}^n$. Consider the enlarged set $\{v_1, \cdots, v_k, e_1, \cdots, e_n\}$, where $e_1, \cdots, e_n$ are the standard basis vectors in $\mathbf{R}^n$. This enlarged set of vectors spans $\mathbf{R}^n$, since the standard basis vectors do. Then we can find a basis of $\mathbf{R}^n$ from the enlarged set by the method in the preceding exercise. (In fact, this basis will always include the vectors $v_1, \cdots, v_k$.)

(a) Find a basis of $\mathbf{R}^4$ that includes the columns v_1, v_2 of

$$V = \begin{pmatrix} 5 & -1 \\ -7 & 2 \\ 1 & -1 \\ 4 & -4 \end{pmatrix}$$

Hint: To construct the enlarged set of vectors, form the augmented matrix $[V\ I]$.

(b) Explain why this method always produces a basis that includes the vectors $v_1, \cdots, v_k$. That is, explain how we know that the reduced row echelon form of $[V\ I]$ will have leading 1's in the first k columns.

(c) Find another basis of $\mathbf{R}^4$ that includes the vectors v_1, v_2 in (a) by simply constructing vectors v_3 and v_4 with random integer entries. Describe how you checked that $\{v_1, v_2, v_3, v_4\}$ is indeed a basis of $\mathbf{R}^4$.

14. A *left inverse* of a matrix A is a matrix B such that $BA = I$. In this exercise, you will explore a method for finding a left inverse. We assume the columns of A are linearly independent. (Otherwise A cannot have a left inverse.) Let $a_1, \cdots, a_k$ denote the columns of A, and let n denote the number of rows of A. Extend $\{a_1, \cdots, a_k\}$ to a basis $\{a_1, \cdots, a_k, \cdots, a_n\}$ of $\mathbf{R}^n$ (see the preceding exercise). Find the inverse C of the matrix with columns $a_1, \cdots, a_k, \cdots, a_n$. Compute CA; from this result you should be able to spot a submatrix of C that is a left inverse of A.

(a) Use this method to find a left inverse of the matrix A in the preceding exercise.

(b) Describe in general how to find a submatrix of C that is a left inverse of A.

A Subspace Can Have Many Different Bases

15. (a) Use the method of Exercise 12 to find a basis for the span of the columns $\{v_1, v_2, v_3, v_4\}$ of the matrix

$$V = \begin{pmatrix} 3 & 4 & -6 & 4 \\ -4 & 0 & 0 & 8 \\ -3 & -2 & 3 & 1 \\ 2 & 6 & -9 & 11 \end{pmatrix}$$

(b) Find all the bases of the span of $\{v_1, v_2, v_3, v_4\}$ in which the basis vectors are taken from the set $\{v_1, v_2, v_3, v_4\}$.

(c) Find all subsets of $\{v_1, v_2, v_3, v_4\}$ that span the same subspace that all four of the vectors span.

16. In this exercise we will explore the following method for testing whether two sets of vectors are bases for the same subspace. Suppose $\{v_1, \cdots, v_k\}$ and $\{w_1, \cdots, w_k\}$ are sets of vectors in $\mathbf{R}^n$. If all the vectors $v_1, \cdots, v_k$ belong to the span of $\{w_1, \cdots, w_k\}$, then each v_i is a linear combination of $w_1, \cdots, w_k$. For example, there are scalars $x_1, \cdots, x_k$ such that

$$v_1 = x_1 w_1 + \cdots + x_k w_k$$

This equation can also be written in matrix form:

$$\mathbf{v}_1 = \begin{bmatrix} \mathbf{w}_1 & \cdots & \mathbf{w}_k \end{bmatrix} \begin{pmatrix} x_1 \\ \vdots \\ x_k \end{pmatrix}$$

We can write each of the remaining vectors $\mathbf{v}_2, \cdots, \mathbf{v}_k$ in a similar way as $\begin{bmatrix} \mathbf{w}_1 & \cdots & \mathbf{w}_k \end{bmatrix}$ times a $k \times 1$ matrix of scalars. Therefore

$$\begin{bmatrix} \mathbf{v}_1 & \cdots & \mathbf{v}_k \end{bmatrix} = \begin{bmatrix} \mathbf{w}_1 & \cdots & \mathbf{w}_k \end{bmatrix} X$$

for some $k \times k$ matrix of scalars X. We can also write this equation as

$$V = WX \tag{4.2}$$

where V is the matrix whose columns are $\mathbf{v}_1, \cdots, \mathbf{v}_k$, and W is the matrix whose columns are $\mathbf{w}_1, \cdots, \mathbf{w}_k$. Therefore $\mathbf{v}_1, \cdots, \mathbf{v}_k$ belong to the span of $\{\mathbf{w}_1, \cdots, \mathbf{w}_k\}$ if and only if the equation $V = WX$ has a solution X. Similarly, $\mathbf{w}_1, \cdots, \mathbf{w}_k$ belong to the span of $\{\mathbf{v}_1, \cdots, \mathbf{v}_k\}$ if and only if $W = VY$ has a solution Y.

(a) Consider the matrices

$$V = \begin{pmatrix} 3 & -3 & 4 \\ 4 & 1 & 1 \\ 1 & -4 & 2 \\ 3 & 1 & 3 \end{pmatrix}, \quad W = \begin{pmatrix} -5 & -1 & 2 \\ 2 & 8 & -11 \\ -3 & -4 & 5 \\ -3 & 4 & -4 \end{pmatrix}$$

Use MATLAB's backslash operator \ to solve the equations $V = WX$ for X and $W = VY$ for Y. Do the columns $\{\mathbf{v}_1, \mathbf{v}_2, \mathbf{v}_3\}$ of V and the columns $\{\mathbf{w}_1, \mathbf{w}_2, \mathbf{w}_3\}$ of W span the same subspace of $\mathbf{R}^4$? (Caution: Since MATLAB's backslash operator \ produces a so-called "least-squares solution", you should check your conclusion by computing the products WX and VY to see if they do equal V and W, respectively.)

(b) Compute the products XY and YX. Explain the relationship you discover between X and Y.

(c) Suppose $\{\mathbf{v}_1, \cdots, \mathbf{v}_k\}$ and $\{\mathbf{w}_1, \cdots, \mathbf{w}_k\}$ are linearly independent sets of vectors in $\mathbf{R}^n$. If $V = WX$ for some matrix X, then, as we showed in (a), $\mathbf{v}_1, \cdots, \mathbf{v}_k$ belong to the span of $\{\mathbf{w}_1, \cdots, \mathbf{w}_k\}$. Prove that the span of $\{\mathbf{v}_1, \cdots, \mathbf{v}_k\}$ and the span of $\{\mathbf{w}_1, \cdots, \mathbf{w}_k\}$ are equal. (Thus, we do not need to prove that $W = VY$ also has a solution when we are working with linearly independent vectors.)

17. There is more than one method for finding a basis for the null space of a matrix A, and different methods may yield different bases.

(a) For the matrix

$$A = \begin{pmatrix} 1 & -3 & -5 & 4 \\ 0 & 1 & 1 & -1 \\ -1 & 3 & 5 & -4 \\ -2 & -4 & 0 & 2 \end{pmatrix}$$

find a basis for its null space two different ways: (i) use the **null** command, and (ii) use the ATLAST command **nulbasis**.

(b) Verify that both bases in (a) do indeed span the same subspace by using the method in the preceding exercise.

4.2 Projects on Vector Space Concepts

1. The Coordinate Game

To play the game, type the ATLAST command **cogame**. The game has five levels, and you choose the level at which you want to play from a menu. The first four levels begin with a graph showing two vectors **u** and **v** in $\mathbf{R}^2$ and a target point O represented by a blue circle. The object of the game is to guess the values of the scalars a and b so that the tip of the vector $a\mathbf{u} + b\mathbf{v}$ will lie in the circle. The first four levels differ only in the choice of the vectors **u** and **v**. Level 5 is a two-person game in which the first player chooses the vectors **u** and **v** and the target point X; the second player must guess the values of a and b.

After playing the game, answer the following questions:

(a) How could you choose the vectors **u** and **v** so that the game is impossible to win?

(b) Find an algebraic method for determining the exact values of a and b whenever the exact coordinates of $\mathbf{u}, \mathbf{v}$, and X are known.

(c) Describe a similar game for vectors in 3-space, assuming we could display vectors in 3-space.

(d) When would it be impossible to win the 3-space game?

2. Understanding the Span

Consider the vectors

$$\mathbf{v}_1 = \begin{pmatrix} 1 \\ 1 \\ 0 \end{pmatrix}, \ \mathbf{v}_2 = \begin{pmatrix} -1 \\ 0 \\ 1 \end{pmatrix}$$

Recall that the span of $\{\mathbf{v}_1, \mathbf{v}_2\}$ is the set of all vectors of the form

$$c_1\mathbf{v}_1 + c_2\mathbf{v}_2 \tag{4.3}$$

where c_1 and c_2 range over all the real numbers. We will denote the span of $\{\mathbf{v}_1, \mathbf{v}_2\}$ by the letter S. Since $\mathbf{v}_1$ and $\mathbf{v}_2$ are in 3-space, we can visualize them; we will draw any vector $[x \ y \ z]^{\mathrm{T}}$ in $\mathbf{R}^3$ as a directed line segment with tail at the origin $(0, 0, 0)$ and head at the point (x, y, z). The purpose of this project is to develop a thorough understanding of the concept of span by exploring this particular span S in detail and visualizing it in $\mathbf{R}^3$.

(a) The vector $\mathbf{v}_1$ is in the span S, as we can see by choosing $c_1 =$ _____ and $c_2 =$ _____ in the expression above.

(b) Similarly, show that $\mathbf{v}_2$ and the zero vector also are in S.

(c) Find any two other vectors in S.

(d) By inspection, not a computation, show that the following two vectors are in S. (Hint: Try some simple values for c_1 and c_2.)

$$\begin{pmatrix} 0 \\ 1 \\ 1 \end{pmatrix}, \begin{pmatrix} 3 \\ 2 \\ -1 \end{pmatrix}$$

(e) Explain why every vector that lies on the same line through the origin as $\mathbf{v}_1$ is in S. (Of course, a similar statement holds for $\mathbf{v}_2$.)

(f) Show that the vector whose head is at the midpoint of the line segment joining the head of $\mathbf{v}_1$ and the head of $\mathbf{v}_2$ is in S.

(g) Show that every vector whose head lies on the line through the heads of $\mathbf{v}_1$ and $\mathbf{v}_2$ is in S. (Hint: The preceding question might give you a hint of how to write such vectors in the form $c_1\mathbf{v}_1 + c_2\mathbf{v}_2$. Or, you might use parametric equations for the line.)

(h) By now you may have guessed that S consists of all the vectors lying in the plane containing the vectors $\mathbf{v}_1$ and $\mathbf{v}_2$. This is quite true. Draw the plane S in an xyz coordinate system. Include in your drawing the parallelogram that connects, in order, the origin, the head of $\mathbf{v}_1$, the head of $\mathbf{v}_1 + \mathbf{v}_2$, the head of $\mathbf{v}_2$, and again the origin.

(i) From your drawing, you should be able to see many vectors that do not lie in S. State one such vector.

(j) The equation of a plane through the origin has the form $ax + by + cz = 0$. Find a, b, and c by substituting the coordinates of $\mathbf{v}_1$ and $\mathbf{v}_2$ into this equation. Check that every vector in S also satisfies this equation.

(k) Check that your answer to (i) is correct by substituting it into the equation you found in (j).

3. Related Dimensions

In this project, you will look for relationships between the dimensions of three different vector spaces associated with a matrix. For an $m \times n$

matrix A, the row space and null space of A are subspaces of $\mathbf{R}^n$ while the column space is a subspace of $\mathbf{R}^m$.

(a) Let

$$A = \begin{pmatrix} 1 & 5 & 1 \\ -2 & 0 & 3 \\ -1 & -3 & 0 \\ 3 & 5 & -2 \end{pmatrix}$$

Use the MATLAB command **rref** to find the reduced row echelon form of A. The echelon form matrix will be denoted by B. From this one computation, you will be able to find a basis for each of the three vector spaces mentioned above and thus find the dimension of each vector space.

(b) Explain why the nonzero rows of B form a basis of the row space of A. State the dimension of the row space of A.

(c) Explain why A and B have the same null space. Solve, by hand, the linear system $B\mathbf{x} = 0$, and from your solution find a basis of the null space of A. State the dimension of the null space.

(d) Let $\mathbf{a}_1, \mathbf{a}_2, \mathbf{a}_3$ denote the columns of A. From the basis you found in (c), show that

$$3\mathbf{a}_1 - \mathbf{a}_2 + 2\mathbf{a}_3 = 0$$

Then explain why $\{\mathbf{a}_1, \mathbf{a}_2\}$ is a basis of the column space of A. State the dimension of the column space.

(e) Let

$$A = \begin{pmatrix} 2 & 1 & -1 & -4 \\ 0 & 2 & 6 & -4 \\ 1 & -1 & -5 & 1 \end{pmatrix}$$

Use the methods above to find bases for the row space, null space, and column space of A. State the dimension of each space.

(f) Construct additional $m \times n$ matrices A of rank r by ATLAST commands of the form

$$A = \mathsf{randint(m,n,5,r)}$$

For each such matrix A, find the three dimensions. Continue constructing matrices A until you see a pattern; then state how the dimensions of the row space, null space, and column space of a matrix are related to one another and to n (the number of columns of A).

(g) Explain why these three dimensions are related in the way you described in (f). (Hint: Look closely at a reduced row echelon form, especially the locations of the pivots.)

4. Write Your Own Quiz

[To the instructor: One way to use this project is to have one team of students design the questions and the answer key and have another team answer the questions.]

(a) In this project, you are asked to make up quiz questions for other students to answer. Also make up an answer key to help you grade their answers.

You have matrices A and B below, where B is the reduced row echelon form of A:

$$A = \begin{pmatrix} 1 & 2 & 3 & 4 \\ 5 & 6 & 7 & 8 \\ 9 & 10 & 11 & 12 \end{pmatrix}, \quad B = \begin{pmatrix} 1 & 0 & -1 & -2 \\ 0 & 1 & 2 & 3 \\ 0 & 0 & 0 & 0 \end{pmatrix}$$

Make up a list of questions that can be answered using these matrices A and B. Here is an example of such a question:

Solve the system of linear equations

$$\begin{aligned} x + 2y + 3z &= 4 \\ 5x + 6y + 7z &= 8 \\ 9x + 10y + 11z &= 12 \end{aligned}$$

Now you make up several more. Here are some suggestions for the kinds of questions you might make up:

i. A question that asks for information about the intersection of a collection of planes.

ii. A question that asks whether or not one vector is in the span of several other vectors.

iii. A question that asks whether or not a set of vectors is linearly independent.

iv. A question that asks for information about the row space of the matrix A.

v. A question that asks for information about the column space of the matrix A.

vi. A question that asks for information about the null space of the matrix A.

(b) Choose a matrix A of your own and find its reduced row echelon form B. Make up a list of questions, similar to those suggested in (a), that can be answered by using these matrices A and B. This time, however, choose a matrix A with a different shape than the matrix in (a); for example, if A is square you can ask questions about the inverse of A.

5. A Concrete Application

Concrete mix, used for such things as sidewalks and building bridges, is comprised of five main materials: cement, water, sand, gravel, and fly ash. By varying the percentages of these materials, mixes of concrete can be produced with differing characteristics. For example, the water to cement ratio affects the strength of the final mix, the sand to gravel ratio affects the "workability" of the mix, and the fly ash to cement ratio affects the durability. Since different jobs require concrete with different characteristics, it is important to be able to produce custom mixes.

You, as the manager of a building supply company, plan to keep on hand three basic mixes of concrete from which you will formulate custom mixes for your customers. The basic mixes have the following characteristics:

	Super-strong (Type S)	All-purpose (Type A)	Long-life (Type L)
cement	20	18	12
water	10	10	10
sand	20	25	15
gravel	10	5	15
fly ash	0	2	8

Each measuring scoop of any mix weighs 60 grams, and the numbers in the table give the breakdown by grams of the components of the mix. Custom mixes are made by combining the three basic mixes. For example, a custom mix might have 10 scoops of Type S, 14 of Type A, and $7\frac{1}{2}$ of type L.

We can represent any mix by a vector (c, w, s, g, f) representing the amounts of cement, water, sand, gravel, and fly ash in the final mix. The basic mixes can also be represented by vectors, say **S**, **A**, and **L**, respectively.

(a) Do you actually need to stock all three basic mixes? Give an explanation that is based on linear algebra concepts.

(b) Describe, using linear algebra concepts, the set of all possible custom mixes in terms of the vectors **S**, **A**, and **L**. Can you get all possible mixes from these three?

(c) A customer requests a custom mix with the following proportions of cement, water, sand, gravel, and fly ash: 16, 10, 21, 9, 4. Find the proportions of type S, type A, and type L mixes needed to create this mix. Is the solution unique? Explain. If the customer wants 5 kilograms (5000 grams) of the custom mix, find the amounts of each of the basic mixes you must use.

(d) Explain why it is always the case that water makes up one-sixth the weight of any custom mix, sand plus gravel makes up half the weight, and cement plus fly ash the remaining third.

(e) By making up a small change in the proportions of the custom mix in (c), invent a custom mix that cannot be created from the three basic mixes. Is it possible to invent a custom mix that satisfies the conditions in (d) but cannot be created from the three basic mixes? If so, invent ones; if not, explain why not.

6. Compare Row Space and Null Space

(a) For the matrix

$$A = \begin{pmatrix} 2 & -4 \\ -3 & 6 \end{pmatrix}$$

find a basis $[x_1 \ x_2]$ of the row space of A and a basis $[y_1 \ y_2]^T$ of the null space of A.

(b) Use the ATLAST command **drawvec** to draw the above row space vector in red and the above null space vector in green:

```
drawvec([x1,x2])
hold on
drawvec([y1,y2],'g')
```

(c) Open a new graphics window by using the command **figure(2)**. Repeat steps (a) and (b) for the matrix

$$A = \begin{pmatrix} -1 & 3 \\ 4 & -12 \end{pmatrix}$$

(d) Open a new graphics window by using **figure(3)**. Repeat steps (a) and (b) for your own 2×2 matrix A of rank 1.

(e) What pattern do you see in (b), (c), and (d)? That is, guess at a relationship that holds between vectors in the row space and vectors in the null space of a 2×2 matrix.

(f) Investigate a similar relationship for 2×3 matrices as follows. For the matrix

$$A = \begin{pmatrix} -2 & 2 & -1 \\ 1 & 4 & -2 \end{pmatrix}$$

verify that the vectors

$$\mathbf{u} = \begin{bmatrix} 1 & 0 & 0 \end{bmatrix} \quad \text{and} \quad \mathbf{v} = \begin{bmatrix} 0 & 1 & -0.5 \end{bmatrix}$$

form a basis for the row space, and that

$$\mathbf{z} = \begin{bmatrix} 0 & 0.5 & 1 \end{bmatrix}^{\mathrm{T}}$$

forms a basis for the null space.

(g) Enter the vectors $\mathbf{z}$, $\mathbf{u}$, and $\mathbf{v}$ from part (f), and use the ATLAST command **protate(z,u,v)** to draw the null space and row space of A. This command draws the vector $\mathbf{z}$ in yellow and the parallelogram determined by the vectors $\mathbf{u}$ and $\mathbf{v}$ in red; then it rotates the entire picture 360° in 3-space. Note that every vector in the parallelogram lies in the row space of A.

```
z = [0,0.5,1]
u = [1,0,0]
v = [0,1,-0.5]
protate(z,u,v)
```

You can slow down the rotation by using a fourth argument that specifies the length in seconds of the pause between frames, as in **protate(z,u,v,0.5)**.

(h) The relationship you described in (e) also holds for 2×3 matrices. If necessary, rephrase your statement in (e) so that it applies to both 2×2 and 2×3 matrices.

(i) There is an algebraic test for checking whether two vectors are orthogonal (i.e., perpendicular). The vectors $\mathbf{u}$ and $\mathbf{v}$ are orthogonal if and only if $\mathbf{u}^T\mathbf{v} = 0$. Use this test to check the correctness of your statement in (e) for all the matrices A in (a), (c), (d), and (f).

(j) For matrices with more than three columns, you cannot check your statement in (e) geometrically. You can, however, use the algebraic test in (i). Let

$$A = \begin{pmatrix} 1 & 0 & 3 & 3 \\ -3 & 1 & -2 & -1 \\ -1 & 1 & 4 & 5 \end{pmatrix}$$

Use the command rref(A) to find a basis for the row space of A and the command nulbasis(A) to find a basis for the null space of A. Use the test in (i) to check the correctness of your statement in (e) for this matrix.

(k) State the dimensions of the row space and null space of the matrices A in (a), (c), (d), (f), and (j). What pattern do you see? That is, guess at a relationship between the dimension of the row space and the dimension of the null space for any m and n matrix. How is this statement connected to the geometric relationship you described in (e)?

7. Column Operations and Null Space Bases

First we review some basic facts about row equivalent matrices and the row echelon form of a matrix. A matrix A is *row equivalent* to a matrix B if some finite sequence of elementary row operations converts A to B. Consequently, there exists a nonsingular matrix P such that $PA = B$. (P is the product of the elementary matrices corresponding to the sequence of elementary row operations.) Row equivalent matrices have the same row space.

The matrix B is a *row echelon form* of A if A is row equivalent to B and B has the following properties: (i) All zero rows of B are at the bottom; and (ii) for each nonzero row, the first nonzero term (called the *leading term*) occurs to the right of the leading term in the preceding row (if there is one). The nonzero rows of such an echelon form of A are linearly independent and hence form a basis of the row space of A. A row echelon form is called a *reduced* row echelon form if (iii) every leading term is 1 and the remaining entries in any column containing a leading term are all 0. Every matrix is row equivalent to a unique reduced row echelon form.

The basic facts about column echelon forms are similar. For example, every matrix A can be converted by a finite sequence of elementary column operations to a unique *reduced column echelon form B*, which has the following properties: (i) All zero columns of B are at the right; (ii) for each nonzero column, the leading term occurs below the leading term in the preceding column (if there is one); and (iii) every leading term is 1 and the remaining entries in any row containing a leading term are all 0. Furthermore, there exists a nonsingular matrix P such that $AP = B$.

(a) By hand, use elementary column operations to convert the following matrix A to column echelon form B (reduced or not):

$$\begin{pmatrix} 3 & 1 & 1 & 0 \\ -1 & -1 & 1 & -2 \\ 2 & 1 & 0 & 1 \end{pmatrix}$$

(b) Verify that the nonzero columns of B are linearly independent; use the definition of linear independence. (This is an instance of the general fact that, when B is a column echelon form of A, the nonzero columns of B form a basis of the column space of A.)

(c) MATLAB can find the reduced column echelon form of a matrix A by finding the reduced row echelon form of A^{T} and then transposing back:

(rref(A'))'

Use this method to find the reduced column echelon form B of

$$A = \begin{pmatrix} 2 & -1 & 3 & 1 & -1 \\ 0 & 0 & -1 & 0 & 1 \\ 1 & -2 & -1 & 1 & 2 \\ 2 & 2 & 1 & 0 & 1 \end{pmatrix}$$

Also, from the echelon form B, find a basis of the column space of A.

(d) In Project 12 of Chapter 2, we learned that if A and B are row equivalent matrices, there is a nonsingular matrix P with the property

$PA = B$. We computed P by computing the reduced row echelon form of $[A\ I]$; we then discovered that the matrix P is a submatrix of the echelon form and it occupies the same position as I does in the matrix $[A\ I]$.

Adapt the method described in the preceding paragraph to find the nonsingular matrix Q with the property $AQ = B$, where A and B are the column equivalent matrices in (c). Hint: As in (c), use the command

$(\mathrm{rref}([A;\mathrm{eye}(5)]'))'$

to compute the reduced column echelon form of

$$\begin{pmatrix} A \\ I \end{pmatrix}$$

(e) One use we can make of the column echelon form is to find a basis for the null space of a matrix. Suppose A is the matrix whose null space we want to find, B is the reduced column echelon form of A, and Q is the nonsingular matrix with the property $AQ = B$. The columns of Q that correspond to the zero columns of B form a basis of the null space of A. Find a basis for the null space of the matrix A in (c). Verify directly from definitions that these vectors are indeed in the null space of A and are linearly independent.

(f) Explain why the method in (e) always produces a basis for the null space of a matrix.

(g) Another use we can make of the column echelon form is to find a solution $\mathbf{x}$ of $A\mathbf{x} = \mathbf{b}$. The first step is to use the method in (d) to find a nonsingular matrix Q such that $[A\ \mathbf{b}]Q$ is the reduced column echelon form of $[A\ \mathbf{b}]$. Look at the columns of Q that correspond to the zero columns of the echelon form. If one of these columns has the form $\begin{pmatrix} \mathbf{x} \\ -1 \end{pmatrix}$ then $A\mathbf{x} = \mathbf{b}$. Explain why this is so. In fact, so long as the last entry of any such column of Q is not zero, we can use this column to find a solution of $A\mathbf{x} = \mathbf{b}$. Explain how. Use this method to solve $A\mathbf{x} = \mathbf{b}$, where A is the matrix in (c) and

$$\mathbf{b} = \mathbf{b}_1 = \begin{pmatrix} -1 \\ 2 \\ 3 \\ 6 \end{pmatrix}, \quad \mathbf{b} = \mathbf{b}_2 = \begin{pmatrix} -1 \\ 2 \\ 3 \\ 5 \end{pmatrix}$$

8. One–Sided Inverses

Only square matrices can have inverses, but nonsquare matrices can have one–sided inverses. That is, if A is an $m \times n$ matrix with $m \neq n$, there may be a matrix B such that $AB = I$ or a matrix C such that $CA = I$. B is called a *right inverse* of A and C a *left inverse*. The purpose of this project is to discover which nonsquare matrices have one–sided inverses and to find a way of computing them using MATLAB.

(a) Generate several nonsquare $m \times n$ matrices of rank r using ATLAST commands of the form $A = \mathsf{randint(m,n,5,r)}$. Try several values of m, n, and r, including some where $m < n$, some where $m > n$, some where r equals the minimum of m and n, and some where r is less than the minimum of m and n. Try to find a right inverse for each of these matrices, by using MATLAB's backslash operator \ to solve the equation $AB = I$ for B. Since the backslash operator always yields a solution, be sure to check the product AB to see if it really is the identity matrix.

(b) For matrices that do have a right inverse, what do you notice about the relation between m, n, and r? Formulate a necessary and sufficient condition for a nonsquare matrix to have a right inverse. (There are many correct formulations; the most useful ones are phrased in terms of linear independence or span of the rows or columns of the matrix.)

(c) State how you would use MATLAB to test whether a matrix has a right inverse and to find a right inverse when it exists; give a worked-out example.

(d) Carry out a similar analysis for the left inverses. (Hint: If a matrix A has a left inverse, what can you say about A^T?)

9. Commute With Me

If A is an $n \times n$ matrix, the *centralizer* of A is the set of all $n \times n$ matrices B that commute with A, i.e., for which $AB = BA$. The purpose of this project is to study the centralizer of a matrix by using a variety of linear algebra concepts.

In particular, we will study the centralizer of

$$A = \begin{pmatrix} 2 & 3 & 0 \\ -5 & -7 & -3 \\ 3 & 4 & 2 \end{pmatrix}$$

That is, we will study the set $\mathcal{V}$ of all 3×3 matrices B that commute with A.

(a) Prove that $\mathcal{V}$ is a subspace of the vector space of all 3×3 matrices.
(b) Prove that if B_1 and B_2 commute with A, so does the product $B_1 B_2$. (Thus $\mathcal{V}$ is not only a subspace but a *subalgebra* of the space of all 3×3 matrices.)
(c) Find all matrices B that commute with A as follows. Write B as a 3×3 matrix of unknowns:

$$B = \begin{pmatrix} r & s & t \\ u & v & w \\ x & y & z \end{pmatrix}$$

We want to find those matrices B such that $AB = BA$; in other words, we want to solve the homogeneous equation $AB - BA = 0$ for the unknown matrix B. We can do this by computing all nine entries of $AB - BA$, setting them equal to zero, and solving for the nine unknowns $r, s, t, u, v, w, x, y, z$. For example, from the $(1,1)$ entry of $AB - BA$, we get the equation $5s - 3t + 3u = 0$. In this way, find by hand a homogeneous linear system of nine equations. Find a basis for the set of solutions of this homogeneous system by using the ATLAST command nulbasis. Then express all the matrices B as linear combinations of these basis solutions.

(d) Clearly I and A both commute with A. So they must be among the matrices B you derived in (c). Show how to obtain I and A from the general form you derived in (c).
(e) What is the dimension of the vector space $\mathcal{V}$?
(f) Find a 3×3 matrix whose centralizer has dimension larger than your answer in (e) but less than 9.
(g) If A is any square matrix of size 2×2 or greater, prove that its centralizer has dimension at least 2. (Hint: Find two linearly independent matrices that commute with A.)

10. Graphs and Incidence Matrices

The *incidence* matrix of a directed graph with m vertices and n edges has m rows and n columns, where the rows correspond to the vertices and the columns to the edges. Each column has one -1 and one 1, and all other entries are 0; the entries -1 and 1 indicate the vertices connected by the

edge that corresponds to the column. Specifically,the row corresponding
to the initial vertex contains the number -1, and the row corresponding
to the terminal vertex contains the number 1. For example, if column six
of an incidence matrix has -1 in row 3 and 1 in row 5, this indicates that
the sixth edge goes from vertex 3 to vertex 5.

Thus, in the following directed graph,

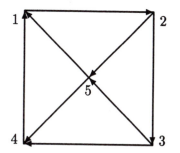

if the edges are written in the order

$$(1\ 2), (2\ 3), (3\ 4), (4\ 1), (2\ 5), (3\ 5), (5\ 4), (5\ 1)$$

where (a, b) denotes the edge from vertex a to vertex b, the corresponding
incidence matrix is

$$\begin{pmatrix} -1 & 0 & 0 & 1 & 0 & 0 & 0 & 1 \\ 1 & -1 & 0 & 0 & -1 & 0 & 0 & 0 \\ 0 & 1 & -1 & 0 & 0 & -1 & 0 & 0 \\ 0 & 0 & 1 & -1 & 0 & 0 & 1 & 0 \\ 0 & 0 & 0 & 0 & 1 & 1 & -1 & -1 \end{pmatrix}$$

In this project you will look for connections between properties of a directed graph and properties of its incidence matrix.

In the questions that follow, you may want to use the ATLAST command
inciden to speed up the process of creating incidence matrices. This command requires a $2 \times n$ matrix whose columns represent the edges. Use
the ATLAST command **nulbasis** to find null space bases.

(a) Create the incidence matrix A above by applying the **inciden** command to the matrix whose columns describe the edges:

$$\begin{pmatrix} 1 & 2 & 3 & 4 & 2 & 3 & 5 & 5 \\ 2 & 3 & 4 & 1 & 5 & 5 & 4 & 1 \end{pmatrix}$$

i. Verify that A has rank 4;

ii. Verify that the vector $[1\ 1\ 1\ 1\ 1]^T$ is a basis for the null space of A^T;

iii. Verify that columns 1,2,3,5 of A form a basis for the column space of A;

iv. Verify that a basis for the null space of A consists of the columns of the matrix

$$\begin{pmatrix} 1 & 0 & 0 & 1 \\ 1 & 1 & -1 & 0 \\ 1 & 0 & -1 & 0 \\ 1 & 0 & 0 & 0 \\ 0 & -1 & 1 & 1 \\ 0 & 1 & 0 & 0 \\ 0 & 0 & 1 & 0 \\ 0 & 0 & 0 & 1 \end{pmatrix}$$

v. Draw the directed graph that gave rise to A, and draw the subgraph corresponding to the above basis of the column space;

vi. For each vector in the above basis of the null space, draw the corresponding set of edges; that is, if the kth entry is $+1$, draw the kth edge, and if the kth entry is -1, draw the kth edge in the opposite direction. For example, the second basis vector corresponds to the edges $(2\ 3), (5\ 2), (3\ 5)$. (A more revealing order is $(2\ 3), (3\ 5), (5\ 2)$.)

(b) For each of the edge matrices below, create the incidence matrix A for the directed graph with the given edges.

i. Find the rank of A;

ii. Find a basis for the null space of A^T;

iii. Find a basis for the column space of A, where all the basis vectors are columns of A;

iv. Find a basis for the null space of A;

v. Draw the directed graph that gave rise to A, and draw the subgraph corresponding to the above basis of the column space;

vi. For each vector in the above basis of the null space, draw the corresponding set of edges.

$$\begin{pmatrix} 1 & 2 & 1 & 4 & 2 & 4 \\ 2 & 3 & 3 & 1 & 4 & 3 \end{pmatrix}$$

$$\begin{pmatrix} 1 & 2 & 3 & 1 & 2 & 3 \\ 2 & 3 & 1 & 4 & 4 & 4 \end{pmatrix}$$

$$\begin{pmatrix} 1 & 2 & 1 & 2 & 3 & 6 & 5 \\ 2 & 3 & 6 & 5 & 4 & 5 & 4 \end{pmatrix}$$

$$\begin{pmatrix} 1 & 1 & 3 & 4 & 4 & 4 & 2 & 3 \\ 2 & 3 & 2 & 1 & 3 & 5 & 5 & 5 \end{pmatrix}$$

(c) From these examples and as many more as you wish to create on your own, describe the patterns you have discovered. That is, describe the connections you see between properties of the incidence matrix and properties of the corresponding directed graph.

(d) Demonstrate your understanding of your discoveries in (c) as follows: using only the picture below of a directed graph, find the rank of its incidence matrix A; a basis for the null space of A^T; a basis for the column space of A; and a basis for the null space of A. (When you do this without matrix computations, you are likely to get different bases for the column space of A and the null space of A than you would using MATLAB.)

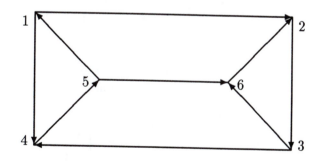

11. A Basis for the Intersection of Two Vector Spaces

Let $V = \mathrm{span}\{\mathbf{v}_1, \mathbf{v}_2\}$, where

$$\mathbf{v}_1 = \begin{pmatrix} -2 \\ 1 \\ 0 \end{pmatrix}, \ \mathbf{v}_2 = \begin{pmatrix} 3 \\ 1 \\ 1 \end{pmatrix}$$

Let $W = \text{span } \{\mathbf{w}_1, \mathbf{w}_2, \mathbf{w}_3\}$, where

$$\mathbf{w}_1 = \begin{pmatrix} 1 \\ -1 \\ -5 \end{pmatrix}, \ \mathbf{w}_2 = \begin{pmatrix} 1 \\ 1 \\ -1 \end{pmatrix}, \ \mathbf{w}_3 = \begin{pmatrix} 0 \\ 1 \\ 2 \end{pmatrix}$$

The purpose of this project is to find a basis for $V \cap W$ (the *intersection* of V and W), by which we mean the set of all vectors that belong to both V and W. For example, the intersection of two nonparallel planes in $\mathbf{R}^3$ is their line of intersection. In discovering how to find a basis of $V \cap W$, you will use many of the basic concepts about vector spaces and systems of equations; thus, this project is suitable for reviewing much of the material covered up to this point of the course.

(a) Use the MATLAB command **rref** to find a basis of V. Determine the dimension of V.

(b) Use **rref** to find a basis of W. Determine the dimension of W.

(c) For each of the following vectors, use the **rref** command to determine whether the vector is in V and whether the vector is in W. Then explain why the vector is or is not in $V \cap W$.

$$\mathbf{u}_1 = \begin{pmatrix} -1 \\ 3 \\ 1 \end{pmatrix}, \ \mathbf{u}_2 = \begin{pmatrix} 4 \\ 8 \\ 4 \end{pmatrix}, \ \mathbf{u}_3 = \begin{pmatrix} -1 \\ 0 \\ 3 \end{pmatrix}$$

(d) Any subspace of $\mathbf{R}^n$ can also be represented as the set of solutions of a system of homogeneous linear equations. For example, if

$$U = \text{Span}\left\{ \begin{pmatrix} 1 \\ 0 \\ 3 \end{pmatrix}, \ \begin{pmatrix} 4 \\ 1 \\ 0 \end{pmatrix} \right\}$$

then U can also be represented by an equation of the form $ax_1 + bx_2 + cx_3 = 0$. Since $[1, 0, 3]$ and $[4, 1, 0]$ are solutions of this equation, we have

$$a + 3c = 0$$
$$4a + b = 0$$

So $[a, b, c]$ is in the null space of the matrix

$$\begin{pmatrix} 1 & 0 & 3 \\ 4 & 1 & 0 \end{pmatrix}$$

Therefore, by applying the **nulbasis** command to this matrix, we will find the coefficients a, b, c. The basis it reports is the vector $[-3, 12, 1]$, and so a linear system that defines U is $-3x_1 + 12x_2 + x_3 = 0$.

Find a homogeneous linear equation that describes the vector space V, by applying the above technique to the vectors v_1, v_2, v_3. Similarly, find a homogeneous linear equation that describes W.

(e) Use the equations in (d) to confirm that the answers you gave in (c) are correct.

(f) Use your answers in (d) to find a system of homogeneous linear equations that describes $V \cap W$.

(g) Use your answer in (f) to find a basis for $V \cap W$. Determine the dimension of $V \cap W$.

(h) Use the method in parts (a) through (g) to find a basis for the intersection of the following two vector spaces:
$V = \text{span}\{v_1, v_2, v_3\}$, where

$$v_1 = \begin{pmatrix} 1 \\ 3 \\ -2 \\ -2 \end{pmatrix}, \quad v_2 = \begin{pmatrix} 4 \\ 0 \\ 1 \\ 4 \end{pmatrix}, \quad v_3 = \begin{pmatrix} -3 \\ -3 \\ 1 \\ 2 \end{pmatrix}$$

$W = \text{span}\{w_1, w_2, w_3, w_4\}$, where

$$w_1 = \begin{pmatrix} 4 \\ 3 \\ 0 \\ -4 \end{pmatrix}, \quad w_2 = \begin{pmatrix} 2 \\ -3 \\ 3 \\ 4 \end{pmatrix}, \quad w_3 = \begin{pmatrix} 1 \\ -2 \\ 2 \\ 0 \end{pmatrix}, \quad w_4 = \begin{pmatrix} 5 \\ 3 \\ 0 \\ 4 \end{pmatrix}$$

Chapter 5

Linear Transformations

5.1 Exercises on Linear Transformations

Linear Transformations from $\mathbf{R}^n$ into $\mathbf{R}^m$ A mapping L from $\mathbf{R}^n$ into $\mathbf{R}^m$ is said to be a *linear transformation* if

$$L(c_1\mathbf{x} + c_2\mathbf{y}) = c_1 L(\mathbf{x}) + c_2 L(\mathbf{y})$$

for all vectors $\mathbf{x}$ and $\mathbf{y}$ in $\mathbf{R}^n$ and for all scalars c_1 and c_2.

1. (a) Let L be a linear transformation from $\mathbf{R}^3$ to $\mathbf{R}^2$ and suppose that $L\left([1,0,2]^{\mathrm{T}}\right) = [2,3]^{\mathrm{T}}$ and $L\left([2,-1,0]^{\mathrm{T}}\right) = [4,-1]^{\mathrm{T}}$. Determine the value of $L\left([-1,1,2]^{\mathrm{T}}\right)$.

   (b) Create a 6×4 matrix of rank 3 by setting U = randint(6,4,5,3). Use the command **nulbasis(U)** to find a vector in the null space of U. This example will not be very interesting if the null space vector has more than one zero component or if its fourth coordinate is 0. If this happens, start over by repeating the **randint** command. Once you have a satisfactory matrix U, set V = randint(5,3,9,3). Let L be a linear transformation that maps each of the first three column vectors of U to the corresponding column vectors of V. Find $L(\mathbf{u}_4)$. (Hint: If you do not see a dependency relation among the column vectors of U, use the null space vector to determine one.)

If L is a linear transformation from $\mathbf{R}^n$ into $\mathbf{R}^m$, then there exists an $m \times n$ matrix A such that

$$L(\mathbf{x}) = A\mathbf{x}$$

for each $\mathbf{x}$ in $\mathbf{R}^n$. We say A is the *matrix representation* of L with respect to the standard basis for $\mathbf{R}^n$.

The *kernel* of L is the set of vectors $\mathbf{x}$ in $\mathbf{R}^n$ that get mapped by L into $\mathbf{0}$. Thus

$$\text{kernel}(L) = \{\mathbf{x} \mid L(\mathbf{x}) = \mathbf{0}\}$$

The *image* of L is the set of vectors $\mathbf{y}$ in $\mathbf{R}^m$ such that $\mathbf{y} = L(\mathbf{x})$ for some $\mathbf{x}$ in $\mathbf{R}^n$. Thus

$$\text{image}(L) = \{\mathbf{y} \mid \mathbf{y} = L(\mathbf{x}), \ \mathbf{x} \in \mathbf{R}^n\}$$

2. The null space and kernel and the column space and image.

(a) Show that if L is a linear transformation from $\mathbf{R}^n$ into $\mathbf{R}^m$ defined by

$$L(\mathbf{x}) = A\mathbf{x}$$

for some $m \times n$ matrix A, then the kernel of L is the null space of A and the image of L is the column space of A.

(b) Set A = randint(6,8,8,5) and let L be the linear transformation from $\mathbf{R}^n$ to $\mathbf{R}^m$ represented by A. Use the **nulbasis** command to find a basis for the kernel of L, and use the **colbasis** command to find a basis for the image of L. What are the dimensions of kernel(L) and image(L)?

3. Dimensions of the kernel and the image.

(a) In each of the following, generate an $m \times n$ matrix A using the command A = randint(m,n,8,r) with the given values of m, n, and r. For each matrix A, determine the dimensions of the kernel and image of the corresponding linear transformation $L(\mathbf{x}) = A\mathbf{x}$. In each case use the **rref** command to compute the reduced row echelon form of A. How many leading 1's does the reduced row echelon form have? How many columns of the reduced row echelon matrix do not have leading 1's?

    i. m = 6, n = 6, r = 4
    ii. m = 7, n = 4, r = 3
    iii. m = 6, n = 5, r = 5

iv. m = 4, n = 7, r = 4

(b) Let L be a linear transformation from $\mathbf{R}^n$ into $\mathbf{R}^m$ with

$$\dim(\text{image}(L)) = r \quad \text{and} \quad \dim(\text{kernel}(L)) = k$$

How do r and k relate to the number of leading 1's in the reduced row echelon form of the matrix representing L? Make a conjecture about the sum $r + k$.

(c) Prove your conjecture from part (b).

Linear Transformations from $\mathbf{R}^n$ into $\mathbf{R}^n$

In many applications it is useful to work with different bases for $\mathbf{R}^n$. Let L be a linear transformation from $\mathbf{R}^n$ into $\mathbf{R}^n$ and let $\{\mathbf{u}_1, \mathbf{u}_2, \ldots, \mathbf{u}_n\}$ be a basis for $\mathbf{R}^n$. Any vector $\mathbf{x} \in \mathbf{R}^n$ can be written as a linear combination

$$\mathbf{x} = c_1\mathbf{u}_1 + c_2\mathbf{u}_2 + \cdots + c_n\mathbf{u}_n$$

or equivalently as a product $\mathbf{x} = U\mathbf{c}$ where

$$U = (\mathbf{u}_1, \mathbf{u}_2, \ldots, \mathbf{u}_n) \quad \text{and} \quad \mathbf{c} = (c_1, c_2, \ldots, c_n)^{\mathrm{T}}$$

The image $L(x)$ can also be represented as a linear combination of the basis vectors

$$L(\mathbf{x}) = d_1\mathbf{u}_1 + d_2\mathbf{u}_2 + \cdots + d_n\mathbf{u}_n$$

or simply as the product $U\mathbf{d}$.

It is possible to represent L by an $n \times n$ matrix in the sense that the coordinate vector $\mathbf{d}$ of the image $L(\mathbf{x})$ can be determined from the coordinate vector $\mathbf{c}$ by a matrix multiplication, i.e., one can find an $n \times n$ matrix B such that $\mathbf{d} = B\mathbf{c}$. The matrix B is said to be the em matrix representation of L with respect to the basis $\{\mathbf{u}_1, \mathbf{u}_2, \ldots, \mathbf{u}_n\}$.

To see how the matrix representation is determined note that

$$L(\mathbf{x}) = c_1L(\mathbf{u}_1) + c_2L(\mathbf{u}_2) + \cdots + c_nL(\mathbf{u}_n)$$

Thus, if we know the effect of L on each of the basis vectors, then it is easy to determine the effect of L on any linear combination of basis vectors. In particular, it can be shown that if

$$L(\mathbf{u}_j) = b_{1j}\mathbf{u}_1 + b_{2j}\mathbf{u}_2 + \cdots + b_{nj}\mathbf{u}_n$$

for $j = 1, \ldots, n$, and B is the matrix whose (i, j) entry is b_{ij}, then B is the matrix representation of L with respect to $\{\mathbf{u}_1, \mathbf{u}_2, \ldots, \mathbf{u}_n\}$.

Alternatively, if we know the matrix representation A with respect to the standard basis $\{\mathbf{e}_1, \mathbf{e}_2, \ldots, \mathbf{e}_n\}$, then it is possible to compute B using the matrices A and U. If

$$L(c_1\mathbf{u}_1 + c_2\mathbf{u}_2 + \cdots + c_n\mathbf{u}_n) = d_1\mathbf{u}_1 + d_2\mathbf{u}_2 + \cdots + d_n\mathbf{u}_n$$

then $L(U\mathbf{c}) = U\mathbf{d}$ and, since A is the standard matrix representation of L, it follows that

$$U\mathbf{d} = AU\mathbf{c}$$

Since U is nonsingular, we can solve for $\mathbf{d}$.

$$\mathbf{d} = U^{-1}AU\mathbf{c}$$

Thus, if we set

$$B = U^{-1}AU$$

then B is the matrix representation of L with respect to $\{\mathbf{u}_1, \mathbf{u}_2, \ldots, \mathbf{u}_n\}$.

4. Set A = randint(4) and consider the linear transformation L defined by $L(\mathbf{x}) = A\mathbf{x}$.

   (a) Set

   $$\mathsf{U} = [1, 1, 1, 1;\ 1, 1, -1, -1;\ 1, -1, 1, -1;\ 1, -1, -1, 1]$$

   and use the **rref** command to verify that the columns of U are linearly independent and hence form a basis for $\mathbf{R}^4$. Determine the matrix B that represents L with respect to this basis.

   (b) Set c = [1:4]′ and x = U*c. Determine the value of $L(\mathbf{x})$ in two ways, by computing $A\mathbf{x}$ and also by setting $\mathbf{d} = B\mathbf{c}$ and computing $U\mathbf{d}$.

5. Set U = triu(ones(5)) and B = toeplitz(1:5). The column vectors of U form a basis of $\mathbf{R}^5$. Let L be the linear transformation represented by the matrix B with respect to $\{\mathbf{u}_1, \mathbf{u}_2, \mathbf{u}_3, \mathbf{u}_4, \mathbf{u}_5\}$.

(a) Set $x = [5:-1:1]'$ and determine its coordinate vector c with respect to $\{u_1, u_2, u_3, u_4, u_5\}$. Compute the coordinate vector d of $L(x)$ with respect to $\{u_1, u_2, u_3, u_4, u_5\}$.

(b) Find the matrix A that represents L with respect to the standard basis for $\mathbf{R}^5$.

(c) Compute $L(x)$ in two ways using the results from parts (a) and (b).

5.2   Projects on Linear Transformations

1. Finding Bases for the Image and Kernel

If L is a linear transformation from $\mathbf{R}^n$ to $\mathbf{R}^m$ then there exists an $m \times n$ matrix A such the $L(\mathbf{x}) = A\mathbf{x}$ for all $\mathbf{x}$ in $\mathbf{R}^n$. The image of L is the column space of A and the kernel of L is the null space of A. One can find bases for these spaces using MATLAB's **orth** and **null** commands. These commands each produce a special type of basis called an *orthonormal basis*. In this project we consider a computationally simpler method to find ordinary bases for the image and kernel of L. This method uses only elementary column operations.

To illustrate how the method works, consider the linear transformation $L(\mathbf{x}) = A\mathbf{x}$ where

$$A = \begin{pmatrix} 1 & 2 & -1 \\ 2 & 5 & 1 \end{pmatrix}$$

The matrix A can be reduced to column echelon form in two steps. In the first step we subtract twice the first column from the second and then add the first column to the third.

$$A = \begin{pmatrix} 1 & 2 & -1 \\ 2 & 5 & 1 \end{pmatrix} \rightarrow A_1 = \begin{pmatrix} 1 & 0 & 0 \\ 2 & 1 & 3 \end{pmatrix}$$

If we denote the column vectors of A by $\mathbf{a}_1$, $\mathbf{a}_2$, and $\mathbf{a}_3$, then the columns of A_1 will be

$$\mathbf{a}_1, \quad -2\mathbf{a}_1 + \mathbf{a}_2, \quad \mathbf{a}_1 + \mathbf{a}_3$$

At the second step we subtract three times the second column of A_1 from the third column:

$$A_1 = \begin{pmatrix} 1 & 0 & 0 \\ 2 & 1 & 3 \end{pmatrix} \rightarrow A_2 = \begin{pmatrix} 1 & 0 & 0 \\ 2 & 1 & 0 \end{pmatrix}$$

The matrix A_2 is in column echelon form. If we express the third column of A_2 in terms of the original columns of A we get

$$\begin{aligned} \mathbf{0} &= \mathbf{a}_1 + \mathbf{a}_3 - 3(-2\mathbf{a}_1 + \mathbf{a}_2) \\ &= 7\mathbf{a}_1 - 3\mathbf{a}_2 + \mathbf{a}_3 \end{aligned}$$

(a) Let A be the matrix in the example. Explain why the first two columns of A form a basis for the image of L. Explain why $\{[7, -3, 1]^T\}$ is a basis for the kernel of L.

If we keep track of the linear combinations of the columns of A that are formed in the reduction to column echelon form, then by noting which combinations get transformed into the zero vector we can find a basis for the kernel of the linear transformation. One way to do this is to augment the matrix A by attaching the identity matrix I as additional rows, i.e., form the block matrix

$$C = \begin{pmatrix} A \\ I \end{pmatrix}$$

Now let us apply the same three column operations to C that were applied to A previously.

$$C = \begin{pmatrix} 1 & 2 & -1 \\ 2 & 5 & 1 \\ \hline 1 & 0 & 0 \\ 0 & 1 & 0 \\ 0 & 0 & 1 \end{pmatrix} \rightarrow \begin{pmatrix} 1 & 0 & 0 \\ 2 & 1 & 3 \\ \hline 1 & -2 & 1 \\ 0 & 1 & 0 \\ 0 & 0 & 1 \end{pmatrix} \rightarrow \begin{pmatrix} 1 & 0 & 0 \\ 2 & 1 & 0 \\ \hline 1 & -2 & 7 \\ 0 & 1 & -3 \\ 0 & 0 & 1 \end{pmatrix}$$

When we apply a column operation to C we are simultaneously applying that same column operation to A and I. As we continue to apply column operations the resulting matrices will have columns that are linear combinations of the columns of C. At the end of the process, the resulting partitioned matrix will have the same linear combination of $\mathbf{a}_1, \mathbf{a}_2, \mathbf{a}_3$ and $\mathbf{e}_1, \mathbf{e}_2, \mathbf{e}_3$ in any column.

The reduction process transformed I into the matrix

$$\begin{pmatrix} 1 & -2 & 7 \\ 0 & 1 & -3 \\ 0 & 0 & 1 \end{pmatrix}$$

It is obvious how to write the third column of this matrix as a linear combination of $\mathbf{e}_1, \mathbf{e}_2, \mathbf{e}_3$:

$$\begin{pmatrix} 7 \\ -3 \\ 1 \end{pmatrix} = 7\mathbf{e}_1 - 3\mathbf{e}_2 + \mathbf{e}_3$$

Since the same column operations were applied to A, the third column of

$$\begin{pmatrix} 1 & 0 & 0 \\ 2 & 1 & 0 \end{pmatrix}$$

must be the same linear combination of $\mathbf{a}_1$, $\mathbf{a}_2$, $\mathbf{a}_3$.

$$\begin{pmatrix} 0 \\ 0 \end{pmatrix} = 7\mathbf{a}_1 - 3\mathbf{a}_2 + \mathbf{a}_3$$

(b) Carry out this process on the matrix

$$A = \begin{pmatrix} 1 & 2 & 0 & 1 & -1 \\ 2 & 1 & 3 & 1 & 0 \\ -1 & 0 & -2 & 0 & 1 \\ 0 & 0 & 0 & 2 & 8 \end{pmatrix}$$

as follows. Enter the matrix A in MATLAB and set $C = [A; \text{eye}(5)]$. Column reducing C is the same as row reducing C^T and then transposing. Set $D = \text{rref}(C')'$. Find a basis for the column space of A and the null space of A by partitioning D into four submatrices and using the column vectors of two of these submatrices.

Consider now the general case where $L(\mathbf{x})$ is a linear transformation from $\mathbf{R}^n$ to $\mathbf{R}^m$ defined by $L(\mathbf{x}) = A\mathbf{x}$. Assume that

$$C = \begin{pmatrix} A \\ I \end{pmatrix}$$

has been reduced using column operations to the form

$$D = \begin{pmatrix} R & O \\ E & K \end{pmatrix}$$

where $(R \ O)$ is a column echelon form of A.

(c) Why do the columns of R belong to the image of L? Why are they linearly independent? Why do they form a basis for the image of L?

(d) Why do the columns of K belong to kernel(L)? Why are they linearly independent? Why do they form a basis for kernel(L)?

2. Computer Graphics

Remember the connect the dots books for children? The same concept can be used to generate computer graphics. The picture is stored as a set of vertices. These vertices are connected by lines to produce the picture. In these exercises, we use this approach to produce the picture of a figure on the screen. We also produce changes in the picture by changing the locations of the vertices and redrawing the figure. Viewing a succession of such drawings can produce the effect of a moving figure as well as other effects such as rotation and enlargement.

How fast the picture will move across the screen depends upon the speed of your computer. It is possible to control the speed of motion using the MATLAB **pause** command. The command **pause(t)** will force a pause of **t** seconds before the execution of the next command. Without the **pause** command the motion may be too fast to see. Unfortunately, MATLAB 4.2 only allows pauses by integer amounts. The command **pause(1)** may slow things up too much, however, a **pause(0)** may slow things up enough. The problem of fine tuning the pauses may only be a temporary inconvenience as it is expected that MATLAB 5.0 will allow pause rates between 0 and 1.

Our goal in these exercises is to investigate techniques that produce the illusion of motion. We consider several basic changes in a figure; changes in the size of the figure, changes in the orientation, changes in location on the screen.

How do matrices enter into the problem? If the changes we want are linear transformations, then the new vertices of our figure can be computed by matrix multiplications.

The basic geometric transformations used in computer graphics are

   (i) the *dilation* and *contraction* of the size of a figure

  (ii) the *reflection* of figures through the axes,

 (iii) the *rotation* of a figure about the origin,

 (iv) the *translation* of a figure

Items (i), (ii), and (iii) are all linear transformations and hence these operations can be accomplished using matrix multiplications. Translations can also be thought of as linear transformations if we use a trick of changing to a different type of coordinate system. While each of the four types of transformations has a simple geometric effect, more complicated effects can be produced by applying combinations of the operations.

(a) **Graphing Figures**

Consider the triangle whose vertices are $(0.5, -1)$, $(0, 1)$, $(-0.5, -1)$.

i. Enter the x and y coordinates of these as MATLAB vectors by setting:

$$x = [-0.5, 0, 0.5, -0.5]$$
$$y = [-1, 1, -1. -1].$$

The coordinates of the first point were entered again in the fourth columns of x and y since we want to connect the third point back to the first. The command plot(x, y) draws this triangle in the graph window. Execute this command to see the plot. The command axis([−2,2,−2,2]) will rescale the axes so that the triangle does not take up the entire graph window. Alternatively, one can store the entire figure in a matrix by storing the x coordinates in the first row and the y coordinates in the second row. Thus to store the triangle in a matrix T set

$$T = [-0.5, 0, 0.5, -0.5; -1.1, -1, -1]$$

or if the vectors x and y have already be entered, then the matrix can be generated by setting T = [x; y]. The triangle can then be plotted using the command

plot(T(1,:),T(2,:))

To view it better set

axis([−2,2,−2,2])

This command should produce the same picture as before since T(1,:) = x and T(2,:) = y.

ii. Consider now the rectangle whose vertices are $(0.5, 1)$, $(-0.5, 1)$, $(-0.5, -0.5)$, $(0.5, -0.5)$. Store the rectangle in a matrix R and plot the graph using the rows of R.

iii. Store the triangle with vertices $(-1, 0.5), (1, 0.5), (0, -1)$ as a matrix $T1$ and the triangle with vertices $(-1, -0.5), (1, -0.5), (0, 1)$ as a triangle $T2$. Plot both figures on the same axis system using the command

$$\text{plot}(T1(1,:),T1(2,:),T2(1,:),T2(2,:))$$

What type of figure results from combining the two triangles?

(b) **Dilation and Contraction**

This exercise shows how to expand or contract the size of a figure. Consider the operation of scalar multiplication by r where $r > 1$. Define $L(\mathbf{x}) = r\mathbf{x}$. It is easy to verify that L is a linear transformation. To determine the matrix for this linear transformations consider the effect of L on the standard basis vectors for $\mathbf{R}^2$. Since

$$L(\mathbf{e_1}) = \begin{pmatrix} r \\ 0 \end{pmatrix} \quad \text{and} \quad L(\mathbf{e_2}) = \begin{pmatrix} 0 \\ r \end{pmatrix}$$

it follows that the matrix representing L is

$$D = \begin{pmatrix} r & 0 \\ 0 & r \end{pmatrix}$$

i. Consider the case where the dilation factor r is 1.25 and enter the corresponding transformation matrix D in the MATLAB command window. Also enter the matrix T representing the triangle in the previous exercise. Plot the triangle and set

    axis([−2,2,−2,2])
    hold on

   Applying L to the vertices of the triangle is equivalent to multiplying D times the columns of T. The matrix $C = DT$ represents the image of the triangle under the transformation L. Compute C and use it to plot the image of the triangle. After plotting the image set **hold off.**

ii. The goal of this problem is to draw the triangle in its original size, then cause it to disappear and redraw it as it looks after it is dilated by the factor 1.25. If this operation is repeated ten times in succession, the triangle will appear to be expanding or moving forward. Because we intend to erase and redraw the figure we will use the command

    p=line(T(1,:),T(2,:),'erasemode','xor');

to draw the graph instead of the **plot** command. After we transform the matrix by setting

T = D*T;

we can erase the original figure and draw the new figure with the command

set(p,'xdata',T(1,:),'ydata',T(2,:));

The following are the commands that produce the effects. Enter them in MATLAB. (You need not enter the comments that follow the % signs.) The **pause(0)** command is used to adjust the speed of the plots so that we can see the graphs before they are erased. This pause rate is not suitable for all computers; you may have to change it depending on the speed of your machine (or even skip the pause statement altogether if you have a very slow computer).

```
clf                            % clear all settings for the plot
p = line(T(1,:),T(2,:),'erasemode','xor');    % plot the triangle
axis([-10 10 -10 10])              % set size of the graph
axis('square')                     % make the display square
figure(gcf)                        % display graphics window
% Adjust the windows on your screen so that both the
% command window and the graphics window show.
hold on                            % hold the current graph
for i = 1:10
    T = D*T;                       % transform the figure
       set(p,'xdata',T(1,:),'ydata',T(2,:)); % erase original figure
                          % and plot the transformed figure
       pause(0) % Adjust this pause rate to suit your computer.
end
hold off
```

iii. Repeat part (ii) for a contraction, that is, scalar multiplication by r, where $0 < r < 1$. What illusion is produced?

(c) Reflections

In this exercise we consider the effect of reflecting a figure about one of the coordinate axes and more generally about any line through the origin. We will use the triangle and rectangle from part (a). The trigonometric identities

$$\cos(\pi - \theta) = -\cos(\theta) \quad \text{and} \quad \sin(\pi - \theta) = \sin(\theta)$$

will also be helpful.

i. Consider the transformation L that reflects a figure through the x-axis. Verify that L is a linear transformation. Determine $L(\mathbf{e}_1)$, $L(\mathbf{e}_2)$ and the matrix A that represents L with respect to the standard basis. Use the matrix A to determine the reflected images of the triangle and rectangle and plot the results.

ii. Suppose that S_w represents a line through the origin that makes an angle w with the x-axis. Let L be the linear transformation that reflects a vector about S_w. Explain why $L(\mathbf{e}_1)$ and $L(\mathbf{e}_2)$ must be unit vectors. What angle will the vector $L(\mathbf{e}_1)$ make with the x-axis? Draw a picture to show geometrically that $L(\mathbf{e}_2)$ will make an angle of $2w - \frac{\pi}{2}$ with the x-axis. Determine the values of $L(\mathbf{e}_1)$ and $L(\mathbf{e}_2)$. If $w = \frac{\pi}{3}$, determine the matrix representing L and use it to reflect the triangle and rectangle.

(d) Rotations

Rotation about the origin is a linear transformation. Intuitively, this is true because rotation is a rigid movement of the plane. It does not change the shapes of figures, merely their orientation with the origin. Consider rotating a vector $\mathbf{v}$ and then multiply it by a scalar c. The same resulting vector is produced by first multiplying $\mathbf{v}$ by c and then rotating $\mathbf{v}$. Similarly, rotation of two vectors followed by their addition produces the same parallelogram and resultant vector that is found by adding the two vectors and then rotating the sum. In the following we will use the triangle from part (a).

i. Let L be the linear transformation that rotates a vector counterclockwise about the origin through the angle θ. Determine $L(\mathbf{e}_1)$, $L(\mathbf{e}_2)$ and the matrix representation of L. The trigonometric identities

$$\cos\left(\frac{\pi}{2} + \theta\right) = -\sin(\theta) \qquad \sin\left(\frac{\pi}{2} + \theta\right) = \cos(\theta)$$

should be helpful.

ii. Determine the matrix representation of the rotation if $\theta = \frac{\pi}{4}$ and plot the image of the triangle for this rotation.

iii. Adapt the procedure developed in part (b) to rotate this triangle
by increments of $\pi/20$ about the origin. Stop when the triangle
is in its original location. If the **pause** command is set properly it
should appear that the triangle is moving around in a circle.

(e) Combinations of Transformations

One can perform a combination of two linear transformations L_1 and
L_2 by first performing the transformation L_1 and then applying the
transformation L_2. If A is the matrix representing L_1 and B is the matrix representing L_2, then the composite transformation is represented
by the product BA.

i. Invent a figure and store its vertices as the columns of a matrix.
Write a program that rotates the figure about the origin, at the
same time expanding its size.

ii. Modify to the program in part (i) to show the figure rotating in a
counterclockwise direction about the origin and expanding at the
same time, then stopping and rotating in the clockwise direction as
it shrinks to its original size. At the end of the program, the figure
should have returned to its original size and original location.

(f) Translations

Reflection and rotation are rigid transformations of the plane. Both
preserve distance. There is one other transformation that preserves
distance. A transformation of the form

$$L(\mathbf{x}) = \mathbf{x} + \mathbf{c}$$

is called a *translation*.

i. Show that if the constant vector $\mathbf{c}$ is nonzero then L is not a linear
transformation.

ii. To see the effect of a translation consider the triangle with vertices
$(1,0), (2,2)$, and $(3,0)$. As in the earlier exercises store the vertices
in a 2×4 matrix T. Plot the graph of the triangle. To apply a
translation by a constant vector $\mathbf{c} = (2,3)^T$, set

```
W(1,:) = T(1,:) + 2;
W(2,:) = T(2,:) + 3;
```

The matrix W will represent the translated triangle. To graph the figure set **axis([0,6,0,6])** and **hold on** and then plot the graph of the translated triangle. While translations are not difficult to do in MATLAB, it is customary in computer graphics to do all transformations as matrix multiplications. Since translations are not linear transformations this presents a problem.

Homogeneous Coordinates.

As was seen in part (f), translations are not linear transformations and consequently cannot be accomplished by matrix multiplication using a 2×2 matrix. Homogeneous coordinates are a trick that can be used to circumvent this problem.

The *homogeneous coordinates* for a point $[x, y]^T$ are $[x, y, 1]^T$. However, when the point represented by the homogeneous coordinates $[x, y, 1]^T$ is plotted, only the x and y are taken into consideration. The 1 is ignored. As we see in the next problems, this trick allows us to produce the translated coordinates for a figure by matrix multiplication. The other types of basic transformations can all be adapted to homogeneous coordinates and interpreted as matrix multiplications.

(g) Dilation, Contraction, Reflection, Rotation

Consider the operation of dilation. Recall that the matrix for expansion by a factor of 1.25 was [1.25, 0; 0, 1.25].

$$\begin{pmatrix} 1.25 & 0 \\ 0 & 1.25 \end{pmatrix} \begin{pmatrix} x \\ y \end{pmatrix} = \begin{pmatrix} 1.25x \\ 1.25y \end{pmatrix}$$

If we change the representation of $[x, y]^T$ to homogeneous coordinates $[x, y, 1]^T$, then we need a 3×3 matrix that will produce the same effect. It is easily seen that matrix

$$M = \begin{pmatrix} 1.25 & 0 & 0 \\ 0 & 1.25 & 0 \\ 0 & 0 & 1 \end{pmatrix}$$

does the trick.

$$\begin{pmatrix} 1.25 & 0 & 0 \\ 0 & 1.25 & 0 \\ 0 & 0 & 1 \end{pmatrix} \begin{pmatrix} x \\ y \\ 1 \end{pmatrix} = \begin{pmatrix} 1.25x \\ 1.25y \\ 1 \end{pmatrix}$$

Use homogeneous coordinates for the vertices of the triangle in the part (f) and form a 3×4 matrix T to represent the triangle.

  i. Use M to perform the dilation on the triangle and plot the results.
 ii. Determine a 3×3 matrix to represent a counterclockwise rotation of $45°$ in the homogeneous coordinate system. Use this matrix to rotate the triangle.
iii. Show that if A is any 2×2 matrix representing linear transformation, then the corresponding matrix for the homogeneous coordinate system is obtained by augmenting A by the row $[0,0,1]$ and the column $[0,0,1]^{\mathrm{T}}$. Hint: use block multiplication to show this works.

(h) Translation

Consider the translation $L(\mathbf{x}) = \mathbf{x} + \mathbf{c}$. In homogeneous coordinates, we have

$$L([x_1, x_2, 1]^{\mathrm{T}}) = [x_1 + c_1, x_2 + c_2, 1]^{\mathrm{T}}$$

We need a matrix M such that

$$M[x_1, x_2, 1]^{\mathrm{T}} = [x_1 + c_1, x_2 + c_2, 1]^{\mathrm{T}}$$

It is easy to verify that the matrix

$$M = \begin{pmatrix} 1 & 0 & c_1 \\ 0 & 1 & c_2 \\ 0 & 0 & 1 \end{pmatrix}$$

has the desired property. Since $L(\mathbf{x}) = M\mathbf{x}$ for each vector $\mathbf{x}$ in the homogeneous coordinate system, it follows that L is a linear transformation for this system.

Repeat (ii) of part (f) using homogeneous coordinates. Represent the triangle by a 3×4 matrix T and the translation by a 3×3 matrix. Compute product representing the translated matrix and plot its graph.

Animations in Homogeneous Coordinates

It is possible to make a figure move across the screen using successive translations just as we made the figures move with other transformations in previous exercises. The ATLAST M-file **movefig** was developed to simplify the process. If F is a matrix representing a figure and M is the matrix of a linear transformation, then the command movefig(F,M,n) will successively apply M to the figure n times. Each time the previous image will be erased before the new image is drawn. Additional input arguments may be used to specify the pause rate and the scaling of the axes in the MATLAB plots.

(i) Let T be the matrix representing the triangle from the previous exercise and let M be the matrix for the translation

$$L(\mathbf{x}) = [x_1 + 0.2, x_2 + 0.1, 1]^{\mathrm{T}}$$

Set z = [0,10,0,10] to specify the axis setting and use the command movefig(T,M,35,0,z) to make the figure move across the screen.

(j) Using homogeneous coordinates let D, R, and S be matrices representing a dilation with $r = 1.1$, a rotation of $\frac{\pi}{30}$ radians, and a reflection of 180°. Describe the operations that are being carried out by each of the following commands:

    movefig(T,D,30,0,[−10,60,−10,40])
    movefig(T,R,60,0)
    movefig(T,S,20,0)

(k) Using the movefig command it is easy to perform a sequence of composite transformations. Using the matrices from the previous parts describe the operations carried out by the commands:

    movefig(T,S*M,60,0.5,[0,15,−10,10])
    movefig(T,D*R,41,0.5,[−100,100,−140,80])
    movefig(T,S*M*D,40,0.5,[−10,250,−100,100])

(l) **Roll Along.**

The ATLAST command rollit(n) will generate the graph of a wheel and cause it to roll a distance n. The motion is simulated using only translations and rotations. The rotations cause the wheel to spin and the translations cause it to move horizontally. However, in order to simulate the spinning properly, the center of wheel must be at the origin. Thus we apply rotations to the wheel in its initial position (centered at the origin). After each rotation the wheel is then translated an appropriate distance horizontally and the resulting figure is plotted. Each successive translation moves the figure further to the right.

i. For $n = 5$, 10, 15, use the ATLAST command **rollit** to simulate a wheel rolling a distance of n units. To see the commands that form the wheel and how the transformations are applied, enter the command **type rollit**. Using this model devise your own M-file **rollback** which simulates the rolling of the wheel from the point $x = n$ back to the origin.

ii. Write an M-file **rollroll** such that **rollroll(n,k)** will simulate a wheel rolling back and forth between $x = 0$ and $x = n$ a total of **k** times.

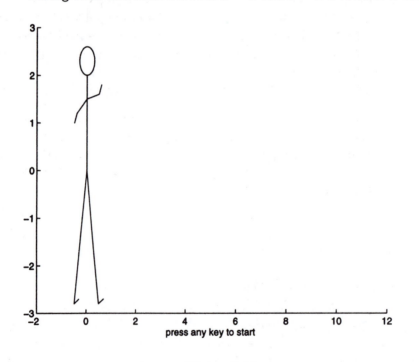

Figure 5.21

(m) **Stick Together.**

The ATLAST command **walk(steps)** will generate a stick figure in the plane (see Figure 5.21). Try the command with **steps = 10**. Press any key to see the figure walk across the screen. The value of **steps** determines how far the figure walks. The motion of the upper body of the figure is generated by a series of horizontal translations. The leg motion is generated by applying rotations about the origin to the legs while they are in their initial position and then translating the rotated legs horizontally to the same x coordinate as the body. It is possible to improve this simulation in a number of ways. We suggest that students work together in teams to modify and enhance the simulation. You may want to start with the file **walk.m**, copy and rename it, and then make whatever changes your team decides on. The following are some suggestions for possible enhancements.

i. Modify the simulation so that the figure swings its arms while walking. To do this you must first translate the arms so that their initial point is the origin. Then apply a rotation to swing the arms and finally apply a translation so that the arms are in the proper position on the translated figure.

ii. Give the poor figure some knees and use two rotations for each leg motion.

iii. Generate a sequence of three or four stick figures (each stored in a separate matrix) so that when plotted and erased in succession the figures simulate one or two steps of motion. To make the figure walk further use a **for** loop and do a translation of the figures in each iteration of the loop.

3. The Transformer Projects

The *Transformer* is an ATLAST utility to help visualize the effects of linear transformations on geometric figures in the plane. The transformer utility features three basic types of linear transformations: rotations, reflections, and diagonal transformations. (It can be shown in general that any linear transformation from $\mathbf{R}^2$ to $\mathbf{R}^2$ can be written as a composite of three of these basic types of transformations.) Figure 5.22 on the next page shows the transformer graphics window after an image has been selected and two transformations have been applied. The first transformation shown in the lower left subplot is a rotation of 45°. The second transformation is a diagonal scaling of the y-coordinates by $\frac{1}{2}$. For information on how to use the transformer, type **help transfor**. The command **transfor** invokes the transformer utility.

Figure 5.22

(a) To Commute or not to Commute

  i. Choose one of the images from the Image pull-down menu at the top
     left of the transformer window. Use the transformation buttons to
     apply a rotation of 60° (in the counterclockwise direction) followed
     by a counterclockwise rotation of 45°. Save the resulting image
     as the Target subplot by clicking on the Freeze button. To see
     the composite transformation matrix, click on the Current Trans
     button. The matrix will be printed in the MATLAB figure window.
     Now click on the Restart button and then apply the same two
     transformations to the initial image but this time in the reverse
     order. Do you end up with the same composite image when you
     reverse the order? Click on the Current Transformation button
     again to see if the composite transformation matrices are the same.
     Do the two rotation matrices commute? Is it possible to represent
     the composite of the two rotations as a single basic transformation
     (either rotation, reflection, or diagonal)? Explain.
  ii. Repeat part (i) except this time start with a counterclockwise ro-
     tation of 135° followed by a diagonal transformation with diagonal
     entries $a = 2$ and $b = 1$.

iii. Repeat parts (i) and (ii) using reflections instead of rotations.

iv. Repeat part (i) twice more, first using a rotation and a reflection and then using two diagonal transformations.

v. Make a list of all possible pairs of the three basic types of transformations and for each pair indicate whether or not the transformations commute. Give either geometric or algebraic justifications for all of your conclusions.

(b) **Inverse Transformations**

If T is a linear transformation from $\mathbf{R}^2$ into $\mathbf{R}^2$ and T is represented by a nonsingular matrix A, then the inverse transformation T^{-1} is given by

$$T^{-1}(\mathbf{x}) = A^{-1}\mathbf{x}$$

for all $\mathbf{x}$ in $\mathbf{R}^2$. In this exercise we will use the transformer to study the effect of applying linear transformations and their inverses to figures in the plane.

i. Let T be a linear transformation represented by a diagonal matrix with diagonal entries $a = 2$ and $b = \frac{1}{2}$. Apply T to one of the images from the transformer image menu. Determine T^{-1} and apply it to the current image. How does the resulting image compare to the initial image. Explain.

ii. Repeat part (i) for the case that the linear transformation T is a rotation of $75°$ in the counterclockwise direction.

iii. Repeat part (i) for the case that the linear transformation T is a reflection about a line through the origin that makes an angle of $60°$ with the x-axis.

iv. Show that in general if T is a linear transformation from $\mathbf{R}^2$ into $\mathbf{R}^2$ and T is represented by a nonsingular matrix A, then

$$T^{-1}(T(\mathbf{x})) = \mathbf{x} \quad \text{and} \quad T(T^{-1}(\mathbf{x})) = \mathbf{x}$$

for all $\mathbf{x}$ in $\mathbf{R}^2$.

v. Select an image from the transformer Image menu and apply a counterclockwise rotation of $135°$ followed by a reflection of $60°$. Determine the inverses of these two transformations and use them to transform the current image back to the initial image. Does it matter which order you use when you apply the inverse transformations? Explain.

vi. Repeat part (v) using a reflection of $45°$ followed by a diagonal transformation with diagonal entries $a = 4$ and $b = 2$.

vii. If T_1 and T_2 are both linear transformations from $\mathbf{R}^2$ into $\mathbf{R}^2$, then the composite transformation $T_2 \circ T_1$ is defined by

$$T_2 \circ T_1(\mathbf{x}) = T_2(T_1(\mathbf{x}))$$

for all $\mathbf{x}$ in $\mathbf{R}^2$. If T_1 and T_2 both have inverses, determine a formula for finding the inverse of $T_2 \circ T_1$.

(c) **Mystery Transformations**

The transformer utility includes a *mystery* game. To play the game, select a level from the Mystery pull-down menu. When you click on any level an unknown linear transformation is applied to the initial image and the resulting transformed image is graphed in the target subplot. The object of the game is to try to guess the mystery transformation. The unknown transformation was constructed using one or more basic transformations. To guess the mystery transformation, apply basic transformations to the initial image and try to get the Current Image subplot to match the Target subplot. If you make incorrect guesses you can use the Undo button to undo the last transformation or the Restart button to start over from the beginning.

i. **Level 1**

At level 1 the mystery transformation is a single basic transformation. If the transformation is a rotation or reflection, then the angle of rotation or reflection will either be a multiple of 45° or a multiple of 60°. If the transformation is diagonal then the diagonal entries of the scaling matrix will either be 1, 2, or 1/2. Play the mystery game at level 1 a few times with each of the three images from the Image menu. How does the image produced by a reflection differ from the image produced by a rotation?

ii. **Level 2**

The level 2 mystery transformations are each composites of two level 1 transformations. One of the two level 1 transformations will always be a diagonal transformation and the other will either be a rotation or a reflection. Play the level 2 mystery game several times with each of the three images from the image menu. After you succeed in matching your Current Image to the Target Window, print the results. (To print the figures in the graph window switch to the MATLAB command window and enter the command **print**.) After some practice you should be able to match the target image using only two transformations. Does the order in which you apply the transformations make a difference? Explain.

iii. **Level 3**

The level 3 mystery transformations are each composites of three level 1 transformations. If the first of the three transformations is either a rotation or reflection, then the second transformation will be diagonal and the third transformation will be either a rotation or reflection. If the first of the transformations is diagonal, then the second will be either a rotation or reflection and the third will be another diagonal transformation. It is usually very difficult to determine a level 3 transformation by trial and error. To help you get started, the first of the three level 1 transformations is printed in the MATLAB command window. Play the level 3 mystery game several times with each of the three images from the image menu. After you succeed in matching your current image to the target window, print the results.

iv. **The Two Person Mystery Game**

Play the mystery game with a partner. For this game, each person takes a turn at constructing a level 2 or level 3 mystery transformation and the other person must find the transformation. The winner is the one who does it with the fewest guesses. Both players should agree on the level of play and the rules for constructing level 1 transformations. Although transformer utility allows you to specify any 2×2 matrix as a linear transformation, it may be impossible to discover an unknown transformation by trial and error unless restrictions are made on what transformations to use. To play the game, send your opponent out of the room while you choose an initial image from the Image menu. The mystery menu is not used to construct the transformation. You must make up the mystery transformation as a combination of level 1 transformations. For example, if you are playing at level 2, make up two level 1 transformations and apply them to the initial image. Click the Freeze and Restart buttons and call your opponent back into the room to start guessing. After your opponent determines the mystery transformation, it is your turn to leave the room.

4. Pitch, Yaw, and Roll

In this project we seek to understand coordinate changes better through applications of the ideas of pitch, yaw, and roll from aircraft maneuvering. Before proceeding, we need to make the meaning of these three terms precise. Roughly speaking, *pitch* is a tilt of the nose of the aircraft up or down; *roll* is banking to the right or left; and *yaw* is the turning of the aircraft to the right or left, horizontally in the absence of any roll. Since each motion is a rotation in a coordinate plane, we need to distinguish between positive and negative angles of rotation. We first agree always to imagine x, y, z coordinate axes such that the x-axis runs the length of the aircraft fuselage, with $x > 0$ at the nose, and the y-axis runs through the wings, with $y > 0$ on the left as we face forward in the aircraft. The positive z-axis points up. Of course, we expect the axis to be mutually perpendicular. Figure 5.23 shows two graphs of a model plane generated by the ATLAST pyr utility file. The first picture shows the initial position of the plane and the second shows the orientation after a pitch, yaw and roll have been applied.

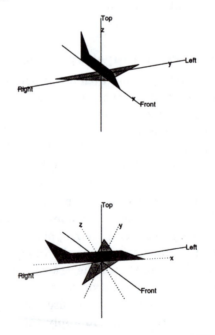

Figure 5.23

For an automobile driver, yaw is the most familiar of the transformations. It is a simple rotation in the xy-plane. If we use w for the angle of yaw, then it is the familiar angle for polar coordinates in the xy-plane, increasing in the counter clockwise direction, that is, in the direction of rotation from the positive x-axis to the positive y-axis. Yaw is similar to turning an automobile, and our point of view is that of the driver. Just as the driver always views the car as going forward, regardless of how it is turning, we will view the positive x-axis always as pointing out the front of the aircraft, regardless of how the aircraft is turning. Likewise, when driving an automobile we think of the combined effect of two turns to the left as equivalent to, but conceptually distinct in operation from, a u-turn. Similarly, roll is a rotation in the yz-plane, and if v is the angle of roll, then v increases when the rotation takes the positive y-axis toward the positive z-axis. Finally, pitch is the rotation of the aircraft fuselage in the xz-plane, but if u is its angle, then u increases when the nose goes down (i.e., the positive z-axis rotates toward the positive x-axis).

(a) **Composite Operations**

Use the ATLAST command **pyr** to generate the graph of a model airplane. Any of the three types of rotations can be accomplished by clicking on the appropriate button and entering the angle of the rotation. The x, y, z axes are drawn in red. These axes change position with the airplane. The initial orientation of the airplane is given by the yellow axes which remain fixed as the plane rotates.

Let P_u stand for the operation of a pitch rotation of the aircraft through the angle of u. Similarly, let R_v and Y_w stand for roll and yaw rotations through angles v and w, respectively. Finally, let $R_v P_u$ stand for pitching by u followed by rolling by v. Notice that in this composite operation, we keep the x-axis pointing out of the nose of the aircraft, even as we are pitching it to a new direction. That is, the coordinate axes travel with the motion of the aircraft, just as you as the pilot would expect. Thus if $u = v = 90°$, under $R_v P_u$ the aircraft ends up with its nose pointing straight down, whereas, if the coordinate axes did not travel with the aircraft, then under $R_v P_u$ its nose would end up pointing in a horizontal direction, but with the wings vertically aligned.

  i. With $u = v = w = 90°$ use the **pyr** model plane to determine the effect of each of the following composite motions, and describe the final orientation relative to its initial orientation.

$$\text{(i) } R_v P_u \qquad\qquad \text{(ii) } P_u R_v \qquad\qquad \text{(iii) } Y_w R_v$$
$$\text{(iv) } R_v Y_w \qquad\qquad \text{(v) } Y_w R_v P_u \qquad\qquad \text{(vi) } Y_w R_v P_u Y_w$$

ii. For different values of u and v investigate the effects of $P_u R_v$ and $R_v P_u$ on your model plane. Are there any values of u and v for which the two composite transformations are the same?

iii. Observe that the inverse operation of P_u is P_{-u}, that is, the combined effect of either followed by the other is to return the aircraft to its original orientation. Likewise R_{-v}, is inverse to R_v, etc. Determine the inverse of each of the following composite operations.

$$\text{(a) } P_u R_v \qquad \text{(b) } R_v P_u \qquad \text{(c) } Y_w R_v \qquad \text{(d) } R_v Y_w$$

(b) Changing Coordinates

In the previous exercise we viewed the aircraft from outside. From such a perspective it may seem a bit strange to think of the coordinate axes as rotating with the aircraft. To aid your intuition in this respect, try to imagine yourself as the pilot of an aircraft in deep space. Consider the apparent effect of a roll thru 90° to the right. It is likely that your experience would not be that the spacecraft was moving, but rather that the stars were rotating. Of course, they would appear to rotate to the left since the spacecraft rotated to the right.

i. Given an object p in the xy-plane at the point $(18, 0, 0)$, what will the coordinates of p be after a yaw of 90°? after a pitch of $-90°$? after a roll of 90°?

ii. Given an object p in the xy-plane at the point $(x_0, y_0, 0)$, what will the coordinates of p be after a yaw of 90°? after a pitch of $-90°$? after a roll of 90°?

iii. Given an object p in the xy-plane at the point $(x_0, y_0, 0)$, what will the coordinates of p be after a yaw of w? after a pitch of u? after a roll of v?

iv. Letting $X = (x_0, y_0, 0)^T$, rewrite your answers to the previous questions in matrix form: $Y_w X$ for the yaw (where now Y_w can denote either the physical operation or a matrix, with context deciding which), and $P_u X$ for the pitch, and $R_v X$ for the roll. Your matrices need only work for the present examples. We will develop them to full generality in the next problems.

v. Given an object p at the point (x_0, y_0, z_0), what will the coordinates of p be after a yaw of 90°? after a pitch of $-90°$? after a roll of 90°?

vi. Given an object p at the point (x_0, y_0, z_0), what will the coordinates of p be after a yaw of w? after a pitch of u? after a roll of v?

vii. Letting $X = (x_0, y_0, z_0)^{\mathrm{T}}$ rewrite your answers to the previous questions in matrix form: $Y_w X$ for the yaw, $P_u X$ for the pitch, and $R_v X$ for the roll.

(c) Using Matrices to Change Coordinates

You should now have fully general matrices P_u, R_v, Y_w from the preceding problems. They can be used to compute the changes in coordinates of an arbitrary point for an arbitrary pitch, yaw, or roll. Their products will give composite rotations.

i. For each of the composite rotations in part (a), compute the corresponding change of coordinates matrix. This is easy enough to do by hand because the matrices have only zeros and ones as entries.

ii. Consider a pitch of $45°$, followed by a roll of $45°$ and then a yaw of $45°$. Form the corresponding product of matrices, $M = Y_w R_v P_u$. If the point X has the coordinates $(0.5000, -0.7071, 0.5000)$ before the rotations, what are the coordinates of X afterwards? Answer the same question for the case where the original coordinates of the point X are $(0.8536, 0.5000 - 0.1464)$.

iii. Consider a pitch of $45°$, followed by a roll of $45°$ and then a yaw of $45°$ and form the matrix $M = Y_w R_v P_u$ as in part (ii). If H represents the half-space which is initially (before the pitch, roll and yaw) above the xy-plane, does the nose of the aircraft point into H after the rotations? (Remember, as a set, H is fixed. Its description by coordinates changes as the aircraft rotates.) How can you use the matrix M to help answer this question? Is it possible to answer the question by examining the value of one of the entries of M? Explain.

iv. How can we determine, after a series of rotations, whether the nose of the aircraft is pointing "forward" with respect to the original axis system. By forward, we mean that the x-coordinate is positive. Answer this question first in the case of a pitch of $1°$ followed by a yaw of $90°$ and then more generally in terms of the entries of a composite matrix M.

v. How can we determine after a series of rotations, whether the wings of the aircraft are tilted more to the right, but with the aircraft still right side up? Before answering this question, we must first phrase the questions precisely. For example, what does it mean for the aircraft to be "still right side up" ? After first clearing up these issues, decide on your criteria for an affirmative answer to the original question. Phrase the criteria in terms of the coordinates of a composite matrix M. Apply your criteria to a roll of 1° followed by a yaw of 90°.

Chapter 6

Orthogonality

6.1   Exercises on Orthogonality

We say that two vectors $\mathbf{u}$ and $\mathbf{v}$ in $\mathbf{R}^n$ are *orthogonal* if their inner product equals zero. The standard inner product for $\mathbf{R}^n$ is the scalar product $\mathbf{u}^T\mathbf{v}$. It is computed in MATLAB using the command u′*v.

As an example, if we set

$$u = [1/\text{sqrt}(2), 1/\text{sqrt}(2)]' \quad \text{and} \quad v = [-1/\text{sqrt}(2), 1/\text{sqrt}(2)]'$$

then clearly $\mathbf{u}$ and $\mathbf{v}$ are orthogonal. MATLAB computes their inner product in finite precision arithmetic, so the computed value on a MacIntosh computer is 2.2362e−17 rather than 0.

The vectors $\mathbf{u}$ and $\mathbf{v}$ given above are examples of *unit vectors*, since they satisfy the conditions $\mathbf{u}^T\mathbf{u} = 1$ and $\mathbf{v}^T\mathbf{v} = 1$ and hence have length 1. We call such unit-length orthogonal vectors "orthonormal".

Let A be the 2×2 matrix whose columns are the vectors $\mathbf{u}$ and $\mathbf{v}$ given above. It follows that

$$A^T A = \begin{pmatrix} \mathbf{u}^T \\ \mathbf{v}^T \end{pmatrix} [\mathbf{u} \ \mathbf{v}] = \begin{pmatrix} \mathbf{u}^T\mathbf{u} & \mathbf{u}^T\mathbf{v} \\ \mathbf{v}^T\mathbf{u} & \mathbf{v}^T\mathbf{v} \end{pmatrix} = I$$

If we compute $A^T A$ in MATLAB and examine the results using **format long**, we see that the computed matrix agrees with the identity matrix to machine precision.

More generally, suppose we want to test whether or not a collection of vectors $\{\mathbf{v}_1, \mathbf{v}_2, \ldots, \mathbf{v}_k\}$ in $\mathbf{R}^n$ is an orthonormal set. We can do this by forming the $n \times k$ matrix V whose column vectors are $\mathbf{v}_1, \mathbf{v}_2, \ldots, \mathbf{v}_k$ and computing the product V^TV. The (i,j) entry of V^TV is $\mathbf{v}_i^T\mathbf{v}_j$. Thus, if the vectors form an orthonormal set, then V^TV must equal the identity matrix I. However, if the vectors $\mathbf{v}_1, \mathbf{v}_2, \ldots, \mathbf{v}_k$ are "close" to orthogonal but not exactly orthogonal, then some of the off diagonal entries of V^TV should be nonzero.

Orthogonal Matrices

If Q is a square matrix and the column vectors of Q form an orthonormal set, then we say that Q is an *orthogonal matrix*.

1. Generate orthogonal matrices A, B, and C by setting

> A = eye(4) − 0.5*ones(4)
> B = cyclic(4)
> C = [54, −82, 18, 6; 54, 18, −82, 6; 18, 6, 6, −98; 62, 54, 54, 18]/100

(a) Verify that A, B, and C are orthogonal matrices.

(b) Use the **inv** command to compute the inverse of each of the three matrices. How are the inverses related to the original matrices?

(c) Compute the products AB, BC, AC and in each case test to see if the product is also an orthogonal matrix.

(d) Set x = randint(4,1) and compute its length using the MATLAB command norm(x). Compute also the lengths of $A\mathbf{x}$, $B\mathbf{x}$, and $C\mathbf{x}$. What do you observe? Prove in general that if Q is an $n \times n$ orthogonal matrix and $\mathbf{x} \in \mathbf{R}^n$, then $\|Q\mathbf{x}\|^2 = \|\mathbf{x}\|^2$ and hence $Q\mathbf{x}$ has the same length as $\mathbf{x}$.

The span of the column vectors of V is the column space of V, which we will denote by $R(V)$. In many applications where a matrix V does not have orthonormal column vectors, it is desirable to find an orthonormal basis for $R(V)$. In the following set of exercises we consider a number of methods for finding an orthonormal basis for the column space of a matrix.

2. Computing Orthonormal Bases

In MATLAB, enter the vectors

$$\mathbf{v}_1 = \begin{pmatrix} -149 \\ 537 \\ -27 \\ 122 \end{pmatrix} \quad \mathbf{v}_2 = \begin{pmatrix} -50 \\ 180 \\ -9 \\ 41 \end{pmatrix} \quad \mathbf{v}_3 = \begin{pmatrix} -154 \\ 546 \\ -25 \\ 129 \end{pmatrix}$$

as columns of a matrix V. Use each of the following methods to find an orthonormal basis for $R(V)$.

(a) The MATLAB command orth(A) returns a matrix whose columns form an orthonormal basis for $R(A)$. Compute an orthonormal basis for $R(V)$ by setting U = orth(V). To test that the column vectors of U form an orthonormal set, compute the product U^TU. To test whether $R(U) = R(V)$, use the MATLAB command

rank(V) == rank([V,U])

Explain why the results of this computation indicate whether or not the two column spaces are equal.

(b) The MATLAB command [Q,R] = qr(A) returns an orthogonal matrix Q and an upper triangular matrix R such that $A = QR$. If A has rank r, then the first r columns of Q form an orthonormal basis for $R(A)$. Use this method to find an orthonormal basis for $R(V)$; that is, set

r = rank(V) and [Q, R] = qr(V)

and then extract the desired columns by setting Q = Q(:,1:r). As in part (a), test to see that the column vectors of Q do indeed form an orthonormal set and whether $R(Q) = R(V)$.

(c) The Gram-Schmidt process can also be used to produce a set of orthonormal vectors. While the method is theoretically important, the classical Gram-Schmidt process as specified in most linear algebra textbooks is prone to numerical instability. In some cases it may give wildly inaccurate results.

There is a modified version of the Gram-Schmidt process that exhibits better numerical stability for floating point computations. The modified version will produce the same orthonormal set of vectors as the classical Gram-Schmidt process if exact arithmetic is used.

The ATLAST M-file **gschmidt** has been developed to perform both versions of the Gram-Schmidt process. To compute an orthonormal basis for $R(V)$ using the modified Gram-Schmidt process, set G = gschmidt(V); to use the classical Gram-Schmidt process, set S = gschmidt(V,1). Compute matrices G and S in this way and then, using MATLAB's **format long**, compute U^TU, Q^TQ, G^TG, S^TS. Which of the four methods does the best job of producing an orthonormal basis? Which method does the worst job?

(d) Use the **randint** command to generate some random $m \times n$ matrices A with integer entries. In each case take $m \geq n$ and use each of the four methods given in Exercise 2 to compute an orthonormal basis for $R(A)$. For each matrix Q produced, evaluate $Q^T Q$ to measure orthonormality. Display the results in **format long**. Compare the results from the four methods.

(e) A matrix A is said to be *ill-conditioned* if solutions to linear systems $A\mathbf{x} = \mathbf{b}$ are sensitive to small changes in the entries of A (such as those produced by roundoff errors). The matrix is *well-conditioned* if the solutions are relatively insensitive to roundoff errors. Generally, the more ill-conditioned a matrix is, the more digits of accuracy you lose when computations are done using floating point arithmetic. For small well-conditioned matrices, all of the four methods discussed should provide orthonormal bases with high accuracy. If, however, the matrix is even moderately ill-conditioned, then the first two methods should be noticeably superior to the Gram-Schmidt methods. This is illustrated in the following example.

The $n \times n$ *Hilbert matrix* H is a matrix whose (i, j) entry is given by

$$h_{ij} = \frac{1}{i + j - 1}$$

The Hilbert matrix can be generated by the MATLAB command H = hilb(n). The Hilbert matrix becomes increasingly ill-conditioned as n increases. For $n = 8$, generate the Hilbert matrix H and compute orthonormal bases for $R(H)$ using each of the four methods. Measure the orthonormality of the computed matrices and compare the results for the four methods. Which methods work the best?

Extending Bases

Once we have found an orthonormal basis $\{\mathbf{u}_1, \mathbf{u}_2, \ldots, \mathbf{u}_k\}$ for a subspace of $\mathbf{R}^n$, we may want to extend that basis to an orthonormal basis for all of $\mathbf{R}^n$. We consider methods for doing this in the following exercise.

3. In Exercise 2, we saw how to find an orthonormal basis for the column space of an $n \times k$ matrix V using either MATLAB's **orth** command or by computing the QR factorization of V. The QR factorization has the added benefit that it automatically gives an extension of the basis to an orthonormal basis for $\mathbf{R}^n$. Form an 8×6 matrix with integer entries and rank 5 by setting

A = randint(8,6,9,5)

(a) Compute the QR factorization of A by setting

[Q,R] = qr(A)

and then set

Q1 = Q(:,1:5)   and   Q2 = Q(:,6:8)

Verify that the column vectors of Q1 form an orthonormal basis for $R(A)$. Since the columns of Q form an orthonormal basis for $\mathbf{R}^8$, it follows that the column vectors q_6, q_7, q_8 of Q2 extend the basis for $R(A)$ to a basis for the entire vector space R^8.

(b) One can find an orthonormal basis for the null space of A^T by setting Z = null(A'). Thus any vector z in $N(A^T)$ can be written in the form

$$z = c_1 z_1 + c_2 z_2 + c_3 z_3 = Z\mathbf{c}$$

Generate vectors $z \in N(A^T)$ and $y \in R(A)$ by setting

c = randint(3, 1)   z = Z * c   x = randint(6, 1)   y = A * x

Compute the value of $y^T z$. Is it equal to 0? Prove in general that if C is an $n \times k$ matrix and $\mathbf{y} \in R(C)$ and $\mathbf{z} \in N(C^T)$, then $\mathbf{y}$ is orthogonal to $\mathbf{z}$. (Hint: If $\mathbf{y} \in R(C)$ then $\mathbf{y} = C\mathbf{x}$ for some $\mathbf{x} \in R^k$.)

(c) Set U = orth(A). Explain why the column vectors of U must be orthogonal to the column vectors of Z. Set

Q = [U Z]

and verify that the column vectors of Q form an orthonormal basis for R^8. Thus the column vectors of Z extend any orthonormal basis for $R(A)$ to an orthonormal basis for $\mathbf{R}^8$.

(d) The matrix Q in part (c) is a square matrix with orthonormal columns. It follows that Q^T must equal Q^{-1} and hence that

$$I = QQ^T = [U\ Z] \begin{pmatrix} U^T \\ Z^T \end{pmatrix} = UU^T + ZZ^T \tag{6.1}$$

Generate a random integer vector in $\mathbf{R}^8$ by setting x = randint(8,1) and then set

$$c = U'*x \qquad y = U*c \qquad d = Z'*x \qquad z = Z*d$$

By construction, $\mathbf{y} \in R(A)$ and $\mathbf{z} \in N(A^T)$. Compute the sum $\mathbf{y} + \mathbf{z}$ and compare the result to $\mathbf{x}$. Explain why the vectors $\mathbf{y}$ and $\mathbf{z}$ constructed this way should add up to $\mathbf{x}$. (Hint: Make use of equation (6.1).) Prove in general that if C is a $n \times k$ matrix and $\mathbf{x} \in \mathbf{R}^n$, then $\mathbf{x}$ can be expressed as a sum $\mathbf{x} = \mathbf{y} + \mathbf{z}$ where $\mathbf{y} \in R(C)$ and $\mathbf{z} \in N(C^T)$.

Least-Squares Problems

Computing a least-squares line

Given a set of points $\{(x_1, y_1), \ldots, (x_k, y_k)\}$, we try to find the line $y = c_1 x + c_2$ which gives the "best fit" in the sense that the sum of the squares of the errors

$$\sum_{i=1}^{k} (y_i - (c_1 x_i + c_2))^2$$

is minimized. The task is to find the values of c_1 and c_2. This is equivalent to finding a least-squares solution of the following matrix equation:

$$\begin{pmatrix} x_1 & 1 \\ x_2 & 1 \\ \vdots & \vdots \\ x_k & 1 \end{pmatrix} \begin{pmatrix} c_1 \\ c_2 \end{pmatrix} = \begin{pmatrix} y_1 \\ y_2 \\ \vdots \\ y_k \end{pmatrix}$$

4. A firm has kept track of the amount spent on advertising and their total sales for each of the first six months of the year. These amounts, reported in thousands of dollars, are summarized in the following table. The x values represent the amounts spent on advertising and the y values are the sales amounts.

x	2.4	3.6	3.6	4.1	4.7	5.3
y	33.8	34.7	35.5	36.0	37.5	38.1

Enter the x and y coordinates in MATLAB as column vectors $\mathbf{x}$ and $\mathbf{y}$. To compute the coefficients of the best linear least-squares fit to the data, set

```
A = [x, ones(size(x))]
c = A\y
```

Use the following commands to plot both the linear function and the data points on the same axis system.

```
t = 2:0.1:6;
z = polyval(c,t);
plot(x,y,'x',t,z)
```

The company plans to spend \$6000 on advertising in July and to keep increasing the amount they spend on advertising by \$500 a month for the rest of the year. Use the linear function to predict their sales for the rest of the year.

Computing a least-squares polynomial

In general, we can find the polynomial of degree n,

$$y = c_1 x^n + c_2 x^{n-1} + \cdots + c_{n-1} x^2 + c_n x + c_{n+1}$$

that gives the best fit of our data points $\{(x_1, y_1), \ldots, (x_k, y_k)\}$. In this case we seek a least-squares solution of the matrix equation

$$\begin{pmatrix} x_1^n & x_1^{n-1} & \cdots & x_1^2 & x_1 & 1 \\ x_2^n & x_2^{n-1} & \cdots & x_2^2 & x_2 & 1 \\ \vdots & \vdots & & \vdots & \vdots & \vdots \\ x_k^n & x_k^{n-1} & \cdots & x_k^2 & x_k & 1 \end{pmatrix} \begin{pmatrix} c_1 \\ c_2 \\ \vdots \\ c_{n-1} \\ c_n \\ c_{n+1} \end{pmatrix} = \begin{pmatrix} y_1 \\ y_2 \\ \vdots \\ y_k \end{pmatrix}$$

5. Let us return to the last exercise and suppose that for the last six months of the year the actual sales for the firm are given in the following table. Again the x values represent the amounts spent on advertising and the y values are the sales amounts.

x	6.0	6.5	7.0	7.5	8.0	8.5
y	38.6	38.7	38.5	37.6	37.2	36.4

Augment the vectors **x** and **y** in Exercise 6 to each include the six additional coordinates and plot the 12 points using the command plot(x,y,'x'). It should be clear from the graph that the linear model in Exercise 6 does not accurately predict sales for the second half of the year. Consequently the linear model should be rejected. As an alternative, consider fitting a quadratic function $y = c_1 x^2 + c_2 x + c_3$ to the entire set of 12 data points. Again we will use MATLAB to compute the best least-squares fit. To generate the coefficient matrix this time, set A = [x.^2, x, ones(size(x))]. As in Exercise 6, compute the vector **c** of least-squares coefficients, and plot the data points (x, y) and the quadratic function whose coefficients are the components of **c**.

Determine a formula for the x value, $xmax$, where the quadratic function assumes its maximum, and calculate this value either by hand or using MATLAB. If the quadratic model is the correct one to use, then $xmax$ is the amount the company should spend on advertising in order to maximize sales.

6.2 Projects on Orthogonality

Computing Least-Squares Circles

Many manufactured parts such as rings, rods, pipes, etc., are circular in shape. Quality control engineers frequently test the parts to see that they meet required specifications. Machines are available to measure coordinates of the points on the perimeter of the manufactured part. The problem then is to determine how close the sampled points come to actually lying on a circle. This can be accomplished by finding the equation of a circle that "best fits" the sampled points and then calculating the distances between the sampled points and the perimeter of the circle.

1. Suppose we are given the following set of data points

x	0.7	3.3	5.6	7.1	6.4	4.4	0.3	-1.1
y	4.0	4.7	4.0	1.3	-1.1	-3.0	-2.5	1.3

The graph of these points is pictured below.

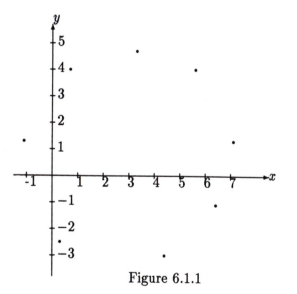

Figure 6.1.1

The points appear to lie roughly on a circle. To measure how close to circular they actually are, let us find the circle that gives the best least-squares fit to the points.

If the circle has radius r and center at (c_1, c_2) then its equation is

$$(x - c_1)^2 + (y - c_2)^2 = r^2.$$

Expanding we get

$$2xc_1 + 2yc_2 + (r^2 - c_1^2 - c_2^2) = x^2 + y^2 \qquad (6.2)$$

By introducing a new variable $c_3 = r^2 - c_1^2 - c_2^2$, we obtain from (6.2) a linear equation for the three unknowns c_1, c_2, and c_3:

$$2xc_1 + 2yc_2 + c_3 = x^2 + y^2.$$

Substituting each of the eight data points (x_i, y_i) into this equation, we obtain a linear system of the form

$$\begin{pmatrix} 2x_1 & 2y_1 & 1 \\ 2x_2 & 2y_2 & 1 \\ \vdots & \vdots & \vdots \\ 2x_8 & 2y_8 & 1 \end{pmatrix} \begin{pmatrix} c_1 \\ c_2 \\ c_3 \end{pmatrix} = \begin{pmatrix} x_1^2 + y_1^2 \\ x_2^2 + y_2^2 \\ \vdots \\ x_8^2 + y_8^2 \end{pmatrix}$$

To solve this using MATLAB, enter the x and y coordinates as column vectors x and y. Construct the coefficient matrix and right-hand side by setting

```
A = [2*x, 2*y, ones(size(x))]
b = x.^2 + y.^2
```

Use MATLAB's \ operation to find the least-squares solution **c** to the system $A\mathbf{c} = \mathbf{b}$. The best fitting circle will have center (c_1, c_2). To determine the radius set

```
r = sqrt(c(3) + c(1)^2 + c(2)^2)
```

To see how well the circle fits the points let us plot the graph of the circle and also plot the original data points. To do this set

```
t = 0:0.1:6.3;
u = c(1) + r*cos(t);
v = c(2) + r*sin(t);
```

and plot the graph using the commands

```
plot(x,y,'x',u,v)
axis('equal')
```

One can measure how close each of the data points is to the center of the circle by setting

$$s = sqrt((x - c(1)).^2 + (y - c(2)).^2)$$

To measure numerically how much the data points deviate from the least-squares circle, set

$$e = s - r$$

The coordinates of **e** show the deviations for each point. Alternatively, to compute a single measure of the overall deviations, set

$$d = norm(e)$$

2. For each of the following data sets, use the method given in Project 1 to determine the center and radius of the circle that gives the best least-squares fit to the data. Assume that you are a quality control engineer and that if any of the data points is more than 0.1 units from the circle, then the part fails to meet specifications. In each case, determine whether or not the sampled points meet the required specifications.

(a)

x	1.2	2.3	3.5
y	5.2	6.4	−1.2

   If all computations for part (a) had been done in exact arithmetic, what would you expect the error to be? Give a geometric explanation of your answer. Why should you always choose more than three sample points?

(b)

x	1.9	−1.1	2.3	4.9
y	2.1	−1.2	−3.9	−0.7

(c) $\dfrac{x}{y}\ \begin{array}{cccccc} 3.3 & 2.0 & -0.3 & -0.8 & 1.4 & 3.0 \\ 1.2 & 3.1 & 2.6 & 0.2 & -1.0 & 0.7 \end{array}$

3. Matrix Models for Measuring Water Contamination

The water quality engineers at a nuclear power plant are interested in measuring the temperature, the level of mercury, and the DO (dissolved oxygen) at a small inlet near the plant. (See Figure 6.2.1.) This inlet has a uniform depth of 30 feet. Fish from the adjacent river enter the inlet, preferring the higher temperatures of the water, and the engineers are interested in protecting them.

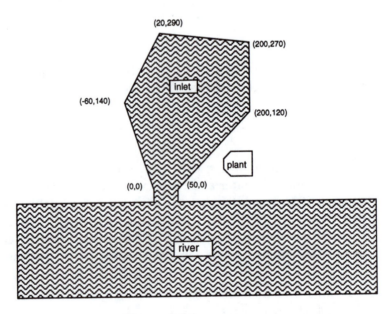

Figure 6.2.1

In June, they installed seven sensors at various points in the inlet and used them to take measurements. The following table lists the (x, y, z) coordinates where the measurements were taken and the recorded values of the temperature (measured in degrees Fahrenheit), the mercury level (measured in micrograms per liter), and the DO (measured in milligrams per liter). The z coordinates represent the depths at which the measurements were taken.

Location			Temp.	Mercury	DO
(10,	10,	0)	66.3	0.279	3.67
(40,	87,	5)	66.7	0.286	3.43
(100,	110,	15)	66	0.286	3.1
(80,	137,	21)	64.8	0.3	3
(147,	220,	25)	66	0.298	2.7
(15,	230,	17)	65	0.317	3.1
(120,	253,	30)	65	0.311	2.6

Table 6.2.1 June Measurements

Using this information, the engineers are looking for a mathematical model which will give them a good estimate of the levels of temperature, mercury, and DO at every point in the inlet. They consulted a statistician who suggested two possible linear models. The first model is given by

$$T(x, y, z) = c_{11} + c_{21}x + c_{31}y + c_{41}z$$
$$M(x, y, z) = c_{12} + c_{22}x + c_{32}y + c_{42}z$$
$$D(x, y, z) = c_{13} + c_{23}x + c_{33}y + c_{43}z$$

where $T(x, y, z)$, $M(x, y, z)$, $D(x, y, z)$, represent the temperature, mercury, and DO level at the point (x, y, z).

The coefficients c_{ij} for this model can be determined by computing least-squares fits to the temperature, mercury, and DO data. For example, to determine the temperature coefficients one must find the least-squares solution to the 7×4 system

$$c_{11} + c_{21}x_i + c_{31}y_i + c_{41}z_i = T(x_i, y_i, z_i) \quad i = 1, \dots, 7$$

(a) Use the MATLAB \ operator to solve for the three sets of coefficients that give the best least-squares fits to the data in Table 6.2.1. Note that each of the three least-squares problems involves the same coefficient matrix A. Consequently, you can combine the three right-hand sides into a 7×3 matrix B and solve all three least-squares problems in one step using the \ operator. The columns of the resulting matrix C = A\B will contain the temperature, mercury, and DO coefficients.

The statistician pointed out that it may be possible that the depth could be ignored in the linear model. This would lead to the alternative model:

$$T(x, y, z) = c_{11} + c_{21}x + c_{31}y$$
$$M(x, y, z) = c_{12} + c_{22}x + c_{32}y$$
$$D(x, y, z) = c_{13} + c_{23}x + c_{33}y$$

(b) The coefficient matrix for this new model is the same as before except that the last column of A is omitted. The new coefficient matrix can be generated in MATLAB by setting A1 = A(:,1:3). Use the \ operator to determine the matrix C1 of least-squares coefficients for this new model.

(c) To test the new model, compute the residuals

R = B − A*C   and   R1 = B − A1*C1

The temperature residuals for the two models are given in the first columns of R and R1. To test whether or not the new model should be used for computing temperatures, set

SS1 = norm(R(:,1))^2
SS2 = norm(R1(:,1))^2
F = 3*(SS2 − SS1)/SS1

The F statistic is a measure of the relative change in the residuals. The factor of 3 in the numerator is the number of degrees of freedom, i.e., the number of data points minus the number of parameters $(7-4)$. The larger F is, the more evidence there is that the two models are not equivalent and that the depth does make a difference.

Generally, whenever we reduce the number of terms in the model we increase the size of the residual. The question is whether this increase is due to chance or to a difference in the models. According to the statistician, if standard assumptions are made about the temperature distribution, then the F statistic will have a certain type of distribution called an F distribution. In particular, if $F > 34.1$, then there is a 99% probability that the increase is not due to chance and that the depth does in fact matter. If $F > 167$ then the probability increases to 99.9%. Compare the computed value of F to these critical values and determine whether or not the alternative model should be used to predict temperatures.

(d) Compute the mercury residuals for the two models and then test whether or not the alternative model should be used to predict mercury levels.

(e) Compute the DO residuals for the two models and test whether or not the alternative model should be used to predict DO levels.

After performing the statistical test, the engineers decided to stay with their original model and not to ignore the depths. The engineers then consulted an environmental biologist on conditions necessary for the survival of fish in the inlet. According to the biologist, the fish will survive as long as the temperature at every point of the inlet is below 73 degrees and the DO level at every point in the inlet is greater than 2.3.

(f) Use the original model to check if the two conditions for fish survival are met. Note that you can use the matrix C to predict the temperature, mercury level, and DO level at any point (x, y, z) by computing $[1, x, y, z]*C$. In particular, it suffices to estimate the temperature and DO levels at the seven points shown in Figure 6.2.1 at depth $z = 0$ and also at depth $z = 30$. We claim that if the conditions for survival of the fish are met at these 14 points, then (according to our linear model) the conditions are met at every point in the inlet. Explain why this claim must be true.

Mercury tends to accumulate in sediments. The State is concerned about high mercury levels and imposes a large fine if the mercury level goes above 4 at the *bottom* of the inlet.

(g) Check the mercury levels at the seven key points at a depth of 30 feet. What is your conclusion? Explain.

The engineers were concerned about the mercury level and decided to take two sets of additional measurements in July and in August. Using the same sensors, they obtained the following measurements:

Location	Mercury (July)	Mercury (August)
(10, 10, 0)	0.29	0.3
(40, 87, 5)	0.298	0.311
(100, 110, 15)	0.299	0.312
(80, 137, 21)	0.31	0.324
(147, 220, 25)	0.312	0.329
(15, 230, 17)	0.331	0.346
(120, 253, 30)	0.325	0.342

Table 6.2.2 July and August Mercury Measurements

(h) Currently we have the June least-squares fit for mercury

$$M(x, y, z) = c_{12} + c_{22}x + c_{32}y + c_{42}z$$

The coefficients in this expression have been computed as the second column of the matrix C. Obtain similar linear models for the mercury levels for the months of July and August.

$$M(x, y, z) = d_1 + d_2x + d_3y + d_4z \quad \text{July}$$
$$M(x, y, z) = e_1 + e_2x + e_3y + e_4z \quad \text{August}$$

Use the same coefficient matrix A as before with the two columns of mercury data from Table 6.2.2 to compute the vectors **d** and **e** of least-squares coefficients for the July and August measurements. Store the June mercury coefficients as a vector **c** by setting **c** = C(:,2).

The statistician was consulted again and this time a dynamic linear model

$$M(t, x, y, z) = r_1 + r_2x + r_3y + r_4z + t(s_1 + s_2x + s_3y + s_4z) \quad (6.3)$$

was recommended. Here the time t is measured in months with $t = 0$ corresponding to the first measurement taken in June. Ideally, for each of the three months $t = 0$, $t = 1$, $t = 2$, we would like the dynamic linear model and the linear model based on that month's data to predict the same values for the seven data coordinates (x_i, y_i, z_i). Ideally then, the r_j and s_j coefficients should satisfy

$$r_1 + r_2x_i + r_3y_i + r_4z_i = c_1 + c_2x_i + c_3y_i + c_4z_i$$
$$r_1 + r_2x_i + r_3y_i + r_4z_i + s_1 + s_2x_i + s_3y_i + s_4z_i = d_1 + d_2x_i + d_3y_i + d_4z_i$$
$$r_1 + r_2x_i + r_3y_i + r_4z_i + 2s_1 + 2s_2x_i + 2s_3y_i + 2s_4z_i = e_1 + e_2x_i + e_3y_i + e_4z_i$$

Since these conditions apply to each of the seven data points, we have a system of 21 equations in 8 unknowns. This overdetermined system can be solved in the least-squares sense.

(i) Show that the system in part (h) can be written in the block matrix form

$$\begin{pmatrix} A & O \\ A & A \\ A & 2A \end{pmatrix} \begin{pmatrix} \mathbf{r} \\ \mathbf{s} \end{pmatrix} = \begin{pmatrix} A\mathbf{c} \\ A\mathbf{d} \\ A\mathbf{e} \end{pmatrix}$$

where $\mathbf{c}, \mathbf{d}, \mathbf{e}$ are the June, July, and August mercury coefficients, and

$$\mathbf{r} = [r_1, r_2, r_3, r_4]' \quad \text{and} \quad \mathbf{s} = [s_1, s_2, s_3, s_4]'$$

are the unknown coefficients in the dynamic model. Form these block matrices in MATLAB and solve the least-squares problem using the \ operation.

(j) The dynamic linear model can be used to predict the level of mercury at the bottom of the inlet for months to come. The point $p = (75, 150, 30)$ is in the middle of the inlet at its bottom. It follows from equation (6.3) that if we set $M(t) = M(t, 75, 150, 30)$, then $M(t)$ is a linear function of the form

$$M(t) = (r_1 + 75r_2 + 150r_3 + 30r_4) + (s_1 + 75s_2 + 150s_3 + 30s_4)t$$
$$= a + bt$$

Use the coefficient vectors $\mathbf{r}$ and $\mathbf{s}$ computed in part (i) to compute the values of a and b. Sketch the graph of the line $M = a + bt$ and solve $a + bt = 4$ to determine the time when the mercury level will reach 4.

Chapter 7

Eigenvalues

7.1 Exercises on Eigenvalues

A Geometric View of Eigenvalues and Eigenvectors

The following exercise presents a geometric interpretation of eigenvalues and eigenvectors for 2×2 matrices. If A is a 2×2 matrix and $\mathbf{x}$ is any unit vector, then generally the directions of $\mathbf{x}$ and $A\mathbf{x}$ will differ. Figure 7.1(a) illustrates the position of $A\mathbf{x}$ relative to $\mathbf{x}$. However, if $\mathbf{x}$ is an eigenvector of A belonging to a real eigenvalue λ, then $A\mathbf{x}$ will be in the same direction as either $\mathbf{x}$ or $-\mathbf{x}$ depending on whether λ is positive or negative (see Figure 7.1(b)).

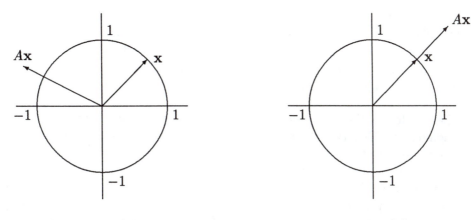

Figure 7.1(a)   Figure 7.1(b)

1. Set $A = [2\ 0;\ 1.5\ 0.5]$. Use the command eigshow(A) to run an animation showing graphically how x and Ax change as x moves around the unit circle. To move the vector x around the circle hold the left button of the mouse down and move the mouse slowly in a counterclockwise circle. Observe the results closely and answer the following questions.

   (a) How many times did x and Ax line up? What is the significance of this in terms of the eigenvalues and eigenvectors of A?

   (b) What type of eigenvalues does A have? Are they real or complex? Are they distinct or multiple? If the eigenvalues are real, how many are positive? negative? zero? Is A nonsingular? Explain.

   (c) Give a rough estimate of the eigenvalues of A.

   (d) Give a rough estimate of the eigenvectors of A. Are they linearly independent? Is A diagonalizable?

   (e) Compute the exact eigenvalues and eigenvectors and compare the results to your estimates.

   (f) Repeat parts (a)–(e) for each of the following matrices

   $$\text{(i)} \begin{pmatrix} -2 & 1 \\ 1 & -2 \end{pmatrix} \qquad \text{(ii)} \begin{pmatrix} -1 & 2 \\ -2 & 4 \end{pmatrix} \qquad \text{(iii)} \begin{pmatrix} 1 & -1 \\ 1 & 1 \end{pmatrix}$$

Special Matrices

This group of exercises involves experimenting with some special matrices in order to determine properties of their eigenvalues and eigenvectors. The *gallery* and *Hadamard* matrices are built-in MATLAB functions. You can obtain information about them using the MATLAB Help facility. The other special matrices are generated by ATLAST M-files. These matrices are described in Appendix B.

2. Let S_n denote the $n \times n$ *sign* matrix. S_n can be generated by the command signmat(n). Generate the matrices S_n for different values of n and in each case compute the eigenvalues of the matrix. Do you see a pattern? Make a conjecture about the values of the eigenvalues of S_n. In particular why must 0 be an eigenvalue and what is its multiplicity? Why must $\lambda = n$ be an eigenvalue? Determine the dimension of the eigenspace corresponding to $\lambda = 0$. Is it possible for S_n to be defective? Explain your answers.

3. Let B_n denote the $n \times n$ *backwards identity* matrix. It can be generated by the command backiden(n).

(a) Generate the matrices B_n for different values of n and in each case compute the eigenvalues of the matrix. Do you see a pattern? Make a conjecture about the values of the eigenvalues of B_n. To prove your conjecture, compute the reduced row echelon forms of $B_n - I$ and $B_n + I$ (consider separately the cases where n is odd and where n is even).

(b) What can you conclude about the multiplicities of the eigenvalues $\lambda = 1$ and $\lambda = -1$; and what can you conclude about the dimensions of the corresponding eigenspaces? Are the matrices B_n diagonalizable? Explain your answers.

4. The *checkerboard* and *anticheckerboard* matrices can be generated by the ATLAST commands **checker** and **achecker**.

(a) Let C_n denote the $n \times n$ *checkerboard* matrix. Generate the matrices C_n for different values of n and in each case compute the eigenvalues of the matrix. Do you see a pattern? Make a conjecture about the values of the eigenvalues of C_n.

(b) If $n > 2$ show that 0 is an eigenvalue of C_n. Determine its multiplicity and the dimension of the corresponding eigenspace.

(c) Repeat parts (a) and (b) in the case that C_n is the $n \times n$ *anticheckerboard* matrix.

5. MATLAB's *gallery* matrices are matrices whose eigenvalues are nice even though they are difficult to compute accurately. Construct the 5×5 gallery matrix by setting A = **gallery(5)**.

(a) Compute the eigenvalues of A by setting e = **eig(A)**. Compute also A^5. What are the eigenvalues of A^5? In general how would you expect the eigenvalues of A^5 to be related to those of A?

(b) Let us now compute the eigenvalues in a different way by finding the roots of the characteristic polynomial. Compute the coefficients of the characteristic polynomial by setting p = **poly(A)**. Use the MATLAB command **roots(p)** to compute the roots of the characteristic polynomial. How do these compare to the eigenvalues computed in part (a)?

(c) Actually, the characteristic polynomial should have integer coefficients. Why? Set p = **round(p)** and compute the roots again. Compare your results to those in parts (a) and (b). Alternatively, if the Symbolic

Toolkit is available one can determine the characteristic polynomial in functional form by setting p= poly(sym(A)). One can then find the roots using the command solve(p).

(d) A matrix B is said to be *nilpotent* if $B^k = O$ for some positive integer k. Prove that the eigenvalues of a nilpotent matrix must all be 0.

(e) Compute the reduced row echelon form of A and determine the nullity of A. Is A diagonalizable? Explain.

6. A real square matrix A is said to be a *square root* of I if $A^2 = I$.

(a) To construct examples of matrices that are square roots of I set
$$G = [0.28, -0.96; -0.96, -0.28]$$
$$H = \text{hadamard}(4)/2$$
Verify that both matrices are square roots of I. Compute the eigenvalues of G and H. Do you notice anything special about the eigenvalues? Show that in general if A is a square root of I and λ is an eigenvalue of A, then λ must equal either -1 or 1.

(b) Compute the traces of the matrices G and H. Show that in general if A is an $n \times n$ square root of I and $\text{trace}(A) = 0$, then n must be even and half of the eigenvalues must be -1 and half must be 1.

(c) Let B_n denote the $n \times n$ backwards identity matrix. Compute B_n^2 for various values of n and verify that each B_n is a square root of I. Using the result from part (b), what can you conclude about the eigenvalues of B_n in the case that n is even? Are your conclusions consistent with those reached in Exercise 3?

The Characteristic Polynomial

In these exercises we explore some of the properties of characteristic polynomials.

7. For each n from 3 to 5 generate a random $n \times n$ matrix A with integer entries by setting A = randint(n). For each matrix A, work through the following exercises.

(a) The characteristic polynomial of A should have integer coefficients. Compute the coefficients by setting p=round(poly(A)). Use the MATLAB commands **trace** and **det** to compute the trace and determinant of A. What relations if any do you observe between the coefficients and the values of the trace and determinant?

(b) Compute the eigenvalues of A by setting $e = eig(A)$. Use the MAT-LAB functions sum and prod to compute the sum and product of the eigenvalues. How are these numbers related to the coefficients of the characteristic polynomial? How do they relate to the trace and determinant of A?

(c) The MATLAB command polyvalm can be used to evaluate a polynomial at a matrix. For example, if $p(x) = x^2 + x + 2$ and B is a square matrix, then $p(B) = B^2 + B + 2I$. If the coefficients of $p(x)$ are stored in the vector $p = [1, 1, 2]$, then the command polyvalm(p,B) will generate the matrix $p(B)$. For each random integer matrix A evaluate it characteristic polynomial p at the matrix A using the command polyvalm(p,A). The observed result is known as the Cayley-Hamilton theorem. State this theorem.

8. Eigenvalues of singular matrices.

(a) For each n from 2 to 7, generate a singular $n \times n$ matrix S by setting
$$S = randint(n,n,10,n-1)$$
In each case determine the characteristic polynomial of S by using the command round(poly(S)).

(b) State and prove a theorem about one of the coefficients and one of the roots of the characteristic polynomial of a singular matrix. If a matrix is nonsingular what can you conclude about the roots of its characteristic equation? Explain.

7.2   Projects on Eigenvalues

The Power Method

1. This project illustrates the geometric effect of applying powers of a 2×2 matrix A to any unit vector **u** and how this is related to the eigenvalues and eigenvectors of A. One can then generalize to the $n \times n$ case where powers of A are used to compute its dominant eigenvalue and the corresponding eigenvector.

   Each unit vector $\mathbf{u} = [u_1, u_2]^T$ can be represented as a point (u_1, u_2) on the unit circle in the plane. One can determine geometrically the effects of a matrix A on all unit vectors by graphing the image of the unit circle under A.

   (a) Consider the matrix

   $$A = \begin{pmatrix} 0.98 & 0.02 \\ 0.20 & 0.80 \end{pmatrix}$$

   Since the rows of A add up to 1, it follows that $\lambda = 1$ is an eigenvalue and $[1, 1]^T$ is an eigenvector. Enter the matrix A and compute the eigenvalues and eigenvectors using the MATLAB command

       [X,D] = **eig**(A)

   Note that the eigenvectors computed by MATLAB all have unit length.

   (b) To see the geometric effect of applying powers of A to unit vectors one can use the ATLAST M-file **powplot**. The command **powplot(A)** will generate a plot of the unit circle and will also plot the eigenvector

   $$\mathbf{z} = \left[\frac{1}{\sqrt{2}}, \frac{1}{\sqrt{2}} \right]^T$$

   belonging to the eigenvalue $t = 1$. It will then successively plot the images of the eigenvector **z** and the unit circle under A^j as j goes from 1 to 25. How does the image of **z** change as j increases? How do the images of the unit circle relate to the eigenspace spanned by **z**?

   (c) To see the effect on a particular unit vector, set

   ```
   x = 2*pi/3;
   u = [cos(x); sin(x)];
   ```

and then use the command **powplot(A,u)**. Make up some additional unit vectors and use the command **powplot** to see their images. Do the images always converge to an eigenvector of $t = 1$?

(d) Set

$$\mathsf{v} = \mathsf{X}(:,2)$$

where X is the matrix of eigenvectors computed in part (a). Use the command **powplot(A,v)** to observe the images. The vector **v** is an eigenvector belonging to the eigenvalue 0.78. Prove that $A^j\mathbf{v}$ must converge to the zero vector as j increases.

(e) Make up some other 2×2 matrices and determine the images of the unit circle under powers of the matrices you constructed. If your matrix has a dominant eigenvalue t (i.e., the other eigenvalue has absolute value less than $|t|$), check to see how the images relate to the eigenspace of t.

(f) Prove that if A is an $n \times n$ diagonalizable matrix with a dominant eigenvalue $t = 1$, then the image of any vector **u** under powers of A will approach a vector **w** in the eigenspace of $t = 1$. Under what conditions will the limit vector **w** be an eigenvector of $t = 1$ and when will **w** be the zero vector? Explain.

(g) If A has a dominant eigenvalue t, then $A^j\mathbf{u}$ will still approach an eigenvector of t as j gets large; however, if $|t| \neq 1$, then the length of $A^j\mathbf{u}$ will approach either 0 or infinity depending upon whether $|t| < 1$ or $|t| > 1$. To avoid this one can scale after each multiplication by A. The following MATLAB commands show how this is done.

$$\mathsf{w} = \mathsf{u}; \quad \mathsf{for\ j} = 1:40 \quad \mathsf{w} = \mathsf{A} * \mathsf{w}; \quad \mathsf{w} = \mathsf{w/norm(w)}; \quad \mathsf{end}, \quad \mathsf{w}$$

Once the eigenvector **w** has been computed, one can approximate the corresponding eigenvalue by setting

$$\mathsf{t} = \mathsf{w'Aw}$$

Why does this formula work? Explain. This method of computing the dominant eigenvalue of a matrix is called the *power method*.

(h) Make up some $n \times n$ matrices and compute the dominant eigenvalue and eigenvector using the power method. Use a number of different values for n and various starting vectors **u**. In each case, compare your results with those obtained using the MATLAB command [X,D] = eig(A).

Eigenvalues of the Alphabet or the Spectrum via Sesame Street

In this subsection we investigate the eigenvalues of special matrices whose shapes resemble letters of the alphabet. These special matrices are described in Appendix B.

2. Let L_n denote the $n \times n$ *letter L* matrix. Generate the matrices L_n for different values of n using the command Lmatrix(n). In each case compute the eigenvectors of the matrix. (Since L_n is lower triangular it is obvious what the eigenvalues should be.) Are the eigenvectors linear independent? Are the *letter L* matrices diagonalizable? Prove your answer.

3. Let n be an odd integer and let H_n denote the $n \times n$ *letter H* matrix. The matrix H_n can be generated using the command Hmatrix(n).

   (a) For $k = 3, 5, 7, 9$, generate H_k and compute its eigenvalues. Make a conjecture about the eigenvalues of H_n.
   (b) What are the rank and nullity of H_n? Explain. Why must $\lambda = 0$ be an eigenvalue of H_n? What is the multiplicity of $\lambda = 0$ and what is the dimension of its eigenspace? Explain.
   (c) Write out the matrices $H_n - 2I$ and $H_n - I$. Without doing any computations how can you tell that $\lambda = 2$ and $\lambda = 1$ are eigenvalues of H_n?
   (d) Are the *letter H* matrices diagonalizable? Explain.

4. Let n be an odd integer and let T_n denote the $n \times n$ *letter T* matrix. The matrix T_n can be generated by the command Tmatrix(n).

   (a) For $k = 3, 5, 7, 9$, generate T_k and compute its eigenvalues. Make a conjecture about the eigenvalues of T_n.
   (b) What are the rank and nullity of T_n? Explain. Why must $\lambda = 0$ be an eigenvalue of T_n? What is the multiplicity of $\lambda = 0$ and what is the dimension of its eigenspace? Explain.
   (c) Write out the matrix $T_n - I$. Without doing any computations how can you tell that $\lambda = 1$ is an eigenvalue of T_n? Use MATLAB to compute the reduced row echelon form of $T_n - I$. What is the dimension of the eigenspace corresponding to $\lambda = 1$?

(d) Are the *letter T* matrices diagonalizable? Explain.

5. Let N_n denote the $n \times n$ *letter N* matrix. It can be generated by the command **Nmatrix(n)**.

   (a) Generate the matrices N_n for different values of $n > 2$ and in each case compute the eigenvalues and eigenvectors. Make a conjecture about the eigenvalues and eigenvectors of N_n.

   (b) Show that for $2 \leq j \leq n - 1$ the vector $\mathbf{e}_j$ (the j-th column vector of the $n \times n$ identity matrix) is an eigenvector of N_n belonging to $\lambda = 1$. Are the *letter N* matrices diagonalizable? Explain.

   (c) Let $p_n(\lambda)$ denote the characteristic polynomial of N_n. Show that $p_2(\lambda) = \lambda^2 - 2\lambda$. Show that

   $$p_k(\lambda) = (1 - \lambda)p_{k-1}(\lambda) \qquad (k > 2)$$

   by expanding $\det(N_n - \lambda I)$ along any inner column. Use this recursion relation to derive a general formula for $p_n(\lambda)$. Does your formula agree with your conjecture about the eigenvalues?

6. In this project we consider the eigenvalues and eigenvectors of powers of the matrices corresponding to letters of the alphabet. If in one of the matrices, say for example the *letter N* matrix, we replace the 1's by other integers, we will still end up with a matrix whose nonzero entries are in the form of the letter N. In this case we say the matrix is in *letter N form*.

   (a) For $k = 2, 3, 4, 5$, compute N_5^k and its eigenvalues. Do the matrices formed by taking powers of N still have the *letter N* form? In general, what will the eigenvalues of N_n^j be? How do the eigenvectors of N_n^j compare to those of N_n? Explain.

   (b) Repeat part (a) for the letter L matrices.

7. **Further Investigations – Group Projects**
   The following are some ideas for further investigations by students. Since these suggestions are somewhat open-ended, it is recommended that students work together in groups to see what they can discover.

   (a) Make up matrices corresponding to other letters of the alphabet such as Z, X or U. Compute the eigenvalues and eigenvectors for your letter matrices and see if any interesting patterns arise. Make conjectures and try to prove that your conjectures are correct.

(b) Replace the 1's in the alphabet matrices by other integers in some systematic way and investigate the eigenvalues and eigenvectors of the resulting matrices. Make conjectures and try to prove that your conjectures are correct.

Eigenvalues of Hermitian Matrices

In Project 2 of Section 2.2 we discovered some elementary properties of symmetric and skew-symmetric matrices. In the following exercise we examine the eigenvalues of these types of matrices. We then consider an important class of matrices called *Hermitian* matrices. In general, if A is a matrix whose entries are complex numbers, then we will use the notation $\overline{A}$ to denote the matrix whose entries are the complex conjugates of the entries of A. One can show that the matrices $\overline{AB}$ and $\overline{A}\,\overline{B}$ are equal by showing that their (i, j) entries are equal.

The conjugate transpose of A, denoted, A^{H}, is defined by $A^{\mathrm{H}} = \overline{A}^{\mathrm{T}}$. Thus A^{H} is the matrix whose (i, j) entry is $\overline{a_{ji}}$, (the complex conjugate of a_{ji}). If A and B are complex matrices and the multiplication AB is possible, then

$$(AB)^{\mathrm{H}} = B^{\mathrm{H}} A^{\mathrm{H}}$$

One can compute the conjugate transpose of a complex matrix A in MATLAB in the same way as one computes the transpose of a real matrix using the $'$ operation. A matrix A is said to be *Hermitian* if $A^{\mathrm{H}} = A$.

8. One can use the command **eigplot(A)** to create a plot of the eigenvalues of A in the complex plane. We can use this command to obtain a geometric view of how the eigenvalues of nonsymmetric, symmetric, and skew symmetric matrices compare. The **eigplot** command can also be used to zoom in on the location of a particular eigenvalue. For a description of how the zoom feature works, type **help eigplot**.

(a) For $n = 2, 3, \ldots, 8$, generate random $n \times n$ matrices A with integer entries and random symmetric matrices B with integer entries by setting:

$$A = \mathsf{randint(n)}$$
$$B = A + A'$$

In each case use the **eigplot** command to observe the graphs of the eigenvalues. Make a conjecture as to the nature of the eigenvalues of a symmetric matrix.

(b) For $n = 2, 3, \ldots, 8$, generate random skew-symmetric $n \times n$ matrices with integer entries by setting

$$C = \mathsf{randint(n)}$$
$$C = C - C'$$

Use the **eigplot** command to plot the eigenvalues of each of the skew-symmetric matrices C. Make a conjecture about the nature of the eigenvalues of a skew-symmetric matrix.

(c) For various values of n compute a symmetric matrix B and a skew-symmetric matrix C (as in parts (a) and (b)) and set $E = B + iC$. Verify that E is Hermitian by computing E^H. In each case use the **eigplot** command to graph the eigenvalues of E. Make a conjecture about the nature of the eigenvalues of an Hermitian matrix.

9. Prove that in general a complex matrix of the form

$$Z = X + iY \qquad (X \text{ and } Y \text{ real})$$

is Hermitian if and only if X is symmetric and Y is skew-symmetric.

10. The *length* of a complex vector $\mathbf{z}$ is defined by

$$\|\mathbf{z}\| = \sqrt{\mathbf{z}^H \mathbf{z}}$$

A vector $\mathbf{u}$ is said to be a *unit vector* if its length is 1, i.e., if $\mathbf{u}^H\mathbf{u} = 1$. If λ is an eigenvalue of a matrix A and $\mathbf{u}$ is a unit eigenvector belonging to λ, then show that

$$\lambda = \mathbf{u}^H A \mathbf{u}$$

Use this equation to prove your conjecture from Exercise 8(c) about the eigenvalues of an Hermitian matrix. (Hint: If λ is a scalar then $\overline{\lambda} = \lambda^H$) Prove your conjecture from Exercise 8(a) about the eigenvalues of a real symmetric matrix.

11. A matrix A is said to be *skew-Hermitian* if $A^H = -A$. Prove that if λ is an eigenvalue of a skew-Hermitian matrix then λ must be purely imaginary, that is, λ must be of the form $\lambda = ai$ where a is a real number.

Eigenvalues of Positive Definite and Positive Semidefinite Matrices

An $n \times n$ symmetric matrix A is said to be *positive definite* if

$$\mathbf{x}^T A \mathbf{x} > 0$$

for every nonzero vector $\mathbf{x}$ in $\mathbf{R}^n$. One way to create a positive definite matrix is to start with a nonsingular matrix B and set $A = B^T B$. If $\mathbf{x}$ is nonzero then $B\mathbf{x} \neq \mathbf{0}$ and it follows that

$$\mathbf{x}^T A \mathbf{x} = \mathbf{x}^T B^T B \mathbf{x} = (B\mathbf{x})^T B \mathbf{x} = \|B\mathbf{x}\|^2 > 0$$

In fact it can be shown that a symmetric matrix A is positive definite if and only if A can be factored into a product $B^T B$ where B is nonsingular.

An $n \times n$ symmetric matrix A is said to be *positive semidefinite* if

$$\mathbf{x}^T A \mathbf{x} \geq 0$$

for every vector $\mathbf{x}$ in $\mathbf{R}^n$ with equality holding for some nonzero vector.

12. For $n = 4, 5, 6, 7, 8$, set

$$B = \text{randint}(n)$$

and use B to construct a positive definite matrix A. Examine the eigenvalues of each matrix A using the **eigplot** command. If an eigenvalue appears to be close to zero, zoom in so that you can estimate it accurately to one decimal place. Make a conjecture about the nature of the eigenvalues of a positive definite matrix.

13. Prove your conjecture from problem 12. Hint: If λ is an eigenvalue of a positive definite matrix A and $\mathbf{u}$ is a unit eigenvector belonging to λ, what is the value of $\mathbf{u}^T A \mathbf{u}$?

14. Show that if B is singular and $A = B^T B$ then A is positive semidefinite. For $n = 4, 5, 6, 7, 8$, set

$$B = \text{randint}(n,n,10,n-1)$$
$$A = B'*B$$

Examine the eigenvalues in the same way as in problem 12. Make a conjecture about the nature of the eigenvalues of a positive semidefinite matrix.

Markov Chains

A *stochastic process* is any sequence of experiments in which the outcome at any stage depends on chance. A *Markov process* is a stochastic process with the following properties:

(i) The set of possible outcomes or states is finite.

(ii) The probability of the next outcome depends only on the previous outcome.

(iii) The probabilities are constant over time.

The following is an example of a Markov process.

> **Voting Trends Example**
> A study has shown that in a certain town 90% of the people who vote Democratic in one election will vote the same way in the next election and the remaining 10% will vote Republican. The study also concluded that 80% of those voting Republican will vote the same way in the next election and the remaining 20% will vote Democratic.

In this example there are two possible outcomes at each stage, a Democratic vote or a Republican vote. The probability of either depends on how one voted at the previous stage. The four probabilities can be summarized in tabular form as follows:

$$
\begin{array}{cc|c}
D & R & \\
\hline
0.90 & 0.20 & D \\
0.10 & 0.80 & R
\end{array}
$$

The first row of the table gives the probabilities that someone who last voted Democrat or Republican will vote Democratic in the next election. The second row gives the probabilities that they will vote Republican. The 2×2 array

$$
A = \begin{pmatrix} 0.90 & 0.20 \\ 0.10 & 0.80 \end{pmatrix}
$$

is called the *transition matrix* for the process. The entries of each column vector of A are probabilities that add up to 1. Such vectors are called *probability vectors*. A square matrix is said to be *stochastic* if each of its column vectors is a probability vector.

If initially there were 5000 Democratic votes and 5000 Republican votes then to predict the voting in the next election set

$$x_0 = \begin{pmatrix} 5000 \\ 5000 \end{pmatrix} \qquad x_1 = Ax_0 = \begin{pmatrix} 5500 \\ 4500 \end{pmatrix}$$

One can predict future election results by setting $x_{n+1} = Ax_n$ for $n = 1, 2, \ldots$. The vectors x_i produced in this manner are referred to as *state vectors* and the sequence of state vectors is called a *Markov chain*. If one divides the entries of the initial state vector x_0 by 10,000 (the total population), the entries of the new vector will represent the proportions of the population in each category. Thus if we take

$$x_0 = \begin{pmatrix} 0.50 \\ 0.50 \end{pmatrix}$$

then x_0 is a probability vector and it is easily seen that each of state vectors in the resulting chain will also be probability vectors.

15. Consider the Markov chain for voting trends example with initial state vector $x_0 = [0.50, 0.50]^T$.

(a) Compute x_1, x_2, x_3 and explain the significance of the entries of each of these vectors. Compute also $A^2 x_0$ and $A^3 x_0$. How do these vectors compare to x_2 and x_3? Explain.

(b) In order to see the long range voting trends, compute the vectors $x_{10}, x_{15}, x_{20}, x_{25}, x_{30}$. Does the sequence of state vectors appear to be converging? In general if the sequence of state vectors converges to a limit vector x, then x is said to be a *steady-state vector* for A.

(c) For each of the following initial vectors, determine whether or not the Markov chain will converge to a steady-state vector. For those that converge, how are the steady-state vectors related?

(i) $x_0 = \begin{pmatrix} 0.25 \\ 0.75 \end{pmatrix}$ \qquad (ii) $x_0 = \begin{pmatrix} 1.00 \\ 0.00 \end{pmatrix}$ \qquad (iii) $x_0 = \begin{pmatrix} 0.00 \\ 1.00 \end{pmatrix}$

(d) A steady-state vector should have the property that $A\mathbf{x} = \mathbf{x}$. Verify that this is indeed the case for any steady state vectors you have found. Interpret this result in terms of eigenvalues and eigenvectors.

(e) If the initial vector is a probability vector, must the steady state vector also be a probability vector? Explain. Compute the eigenvalues and eigenvectors of the transition matrix A. In view of part (c), how many steady state vectors are possible?

16. For $n = 2, 3, 4, 5, 6$, generate an $n \times n$ stochastic matrix by setting

$$A = \mathsf{randstoc(n)}$$

In each case compute the eigenvalues of A and compute the sum of the rows of $A - I$ by using the MATLAB command $\mathsf{sum(A \text{ - } eye(n))}$.

(a) Make a conjecture about the sum of the row vectors of $A - I$ if A is stochastic.

(b) Make and prove a conjecture about one of the eigenvalues of a stochastic matrix.

17. Let A be an $n \times n$ stochastic matrix with eigenvalues $\lambda_1, \lambda_2, \ldots, \lambda_n$ and corresponding eigenvectors $\mathbf{v}_1, \mathbf{v}_2, \ldots, \mathbf{v}_n$.

(a) If the initial state vector can be written as a linear combination

$$\mathbf{x}_0 = c_1\mathbf{v}_1 + c_2\mathbf{v}_2 + \cdots + c_n\mathbf{v}_n$$

determine $\mathbf{x}_1$ and $\mathbf{x}_2$ as linear combinations of $\mathbf{v}_1, \mathbf{v}_2, \ldots, \mathbf{v}_n$. Derive a general formula for expressing $\mathbf{x}_k$ as a linear combination of $\mathbf{v}_1, \mathbf{v}_2, \ldots, \mathbf{v}_n$.

(b) Assume A has distinct eigenvalues. If $\lambda_1 = 1$ and $|\lambda_j| < 1$ for $j = 2, 3, \ldots, n$, show that the chain will have a steady-state vector and that it will be in the direction of $\mathbf{v}_1$.

18. In some situations the transition matrix is not a stochastic matrix. For example, let's suppose the population dynamics of the speckled phoenix have been studied carefully in order to make decisions about clear-cutting in its habitat. There are three components of the population to consider:

juvenile females (denoted by j), subadult females (denoted by s), and mature females (denoted by m). We count only females because the phoenix mates for life, so we assume a one-to-one ratio of females to males. The transition matrix determined by the study is

$$
A = \begin{array}{c} \begin{array}{ccc} j & s & m \end{array} \\ \left(\begin{array}{ccc} 0 & 0 & 0.4 \\ 0.2 & 0 & 0 \\ 0 & 0.7 & 0.9 \end{array} \right) \begin{array}{c} j \\ s \\ m \end{array} \end{array}
$$

The entries of A represent the fraction of the population that moves to the next stage in one time period except for the case of the (1,3) entry. In this case $a_{13} = 0.4$ means that juveniles are produced at a rate of 40% of the mature population.

(a) Explain how the eigenvalues of A can be used to predict what will happen to the speckled phoenix population in the long run. Compute the eigenvalues of A and use them to predict what will happen to the speckled phoenix.

(b) Suppose the (2,1) entry of A is 0.4 instead of 0.2. This would mean 40% of the juveniles survive to become subadults. What would happen to the speckled phoenix under these conditions?

Gerschgorin Circles

In 1931 S. Gerschgorin proved two remarkable theorems which have a wide range of applications. The theorems allow one to obtain useful information about the location in the complex plane of the eigenvalues of a matrix without actually having to compute the eigenvalues.

The First Gerschgorin Circle Theorem

Every eigenvalue of an $n \times n$ matrix A lies in at least one of the circles having centers $c_i = a_{ii}$ and radii $r_i = \sum_{j \neq i} |a_{ij}|$, $i = 1, \ldots, n$.

The Second Gerschgorin Circle Theorem

If k of the Gerschgorin circles of A are disjoint from the other circles, then there are k eigenvalues of A lying inside the union of those circles.

One can plot the Gerschgorin circles of a matrix using the ATLAST M-file **gersch**. The command **gersch(A)** will plot the Gerschgorin circles of A. The command **gersch(A,1)** will plot both the Gerschgorin circles and the eigenvalues of A.

19. For each of the following triangular matrices, identify the centers and radii of the Gerschgorin circles and identify the eigenvalues of the matrix. Plot the circles and the eigenvalues using the **gersch** command.

(a) $\begin{pmatrix} 2 & -1 & 0 \\ 0 & -2 & 2 \\ 0 & 0 & -5 \end{pmatrix}$ (b) $\begin{pmatrix} 2 & -1 & 1 \\ 0 & -2 & 4 \\ 0 & 0 & -5 \end{pmatrix}$

20. For each of the following matrices, identify the centers and radii of the Gerschgorin circles. Plot the circles and the eigenvalues using the **gersch** command. Must each circle contain an eigenvalue? Explain.

(a) $\begin{pmatrix} 10 & 2 & 0 \\ -1 & 3 & -5 \\ 0 & 1 & -1 \end{pmatrix}$ (b) $\begin{pmatrix} 6 & 2 & i \\ -1 & 3-2i & -4+3i \\ 0 & 1 & -2i \end{pmatrix}$

21. The First Gerschgorin Circle Theorem can be proved in three steps.

(a) Show that if λ is an eigenvalue of an $n \times n$ matrix A with corresponding eigenvector **x**, then the relationship

$$(\lambda - a_{ii})x_i = \sum_{j \neq i} a_{ij}x_j$$

holds for every $i = 1, \ldots, n$.

(b) Select the index i of part (a) so that $|x_i| = \max_{1 \leq j \leq n} |x_j|$. Thus $|x_j|/|x_i| \leq 1$ for each $j = 1, \ldots, n$. Use the relationship from part (a) to show that for this index i

$$|\lambda - a_{ii}| \leq \sum_{j \neq i} |a_{ij}|$$

(c) Show that the result of part (b) is equivalent to the given statement of the First Gerschgorin Circle Theorem.

22. Given

$$A = \begin{pmatrix} -5 & -3 & 0 & 1 \\ -1 & 10 & 0 & 0 \\ 1 & 2 & -10 & 0 \\ 1 & 1 & 1 & 20 \end{pmatrix}$$

(a) Use the command **gersch(A,0,'r')** to plot the Gerschgorin circles of A. The third argument 'r' specifies the color of the plot. Examine the plot and use Gerschgorin's second theorem to determine how many eigenvalues of A are in each of the three regions defined by the circles.

(b) Since the entries of A are all real, any complex eigenvalues must occur in conjugate pairs. Explain why the circles centered at (10,0) and (20,0) must each contain exactly one real eigenvalue of A. In general what can you conclude about the eigenvalues of a real matrix if all of its Gerschgorin circles are disjoint?

(c) Since A^T has the same eigenvalues as A, one can obtain more information about the location of the eigenvalues of A by examining the Gerschgorin circles of A^T. Plot the Gerschgorin circles of A^T on the same figure as the plot in part (a) by using the commands

    hold on
    gersch(A',0,'y')

What can you conclude about the nature and location of the eigenvalues of A from examining the resulting figure? Explain. (Remember to unfreeze the plot with the command **hold off.**)

A proof of the Second Gerschgorin Circle Theorem depends on the fact that the zeros of a polynomial are continuous functions of the coefficients of the polynomial. Since the coefficients of the characteristic polynomial of a matrix A are themselves continuous functions of the entries of A, it follows that the eigenvalues of a matrix, being the zeros of the characteristic polynomial, are continuous functions of the entries of the matrix. The proof of the second theorem exploits this fact by examining the behavior of the eigenvalues of the family of matrices $A(t) = D + tE$ as the scalar t ranges from 0 to 1; here D is a diagonal matrix whose entries are the diagonal entries of A, and $E = A - D$ is the matrix whose off-diagonal entries are the remaining entries of A and whose diagonal entries are all 0.

23. Assume A is an arbitrary $n \times n$ matrix and $A(t) = D + tE$.

(a) Describe the matrix $A(0)$. What will its Gerschgorin circles look like?

(b) What is the matrix $A(1)$ and what will its Gerschgorin circles look like?

(c) If $0 < t < 1$, identify the centers and radii of the Gerschgorin circles of $A(t)$. How do the Gerschgorin circles of $A(t)$ compare to the Gerschgorin circles of A?

24. To illustrate the idea behind the proof of the second Gerschgorin theorem, consider the following example. Let

$$A = \begin{pmatrix} 12 & 2 & 0 \\ -1 & 3 & -5 \\ 0 & 1 & -1 \end{pmatrix}$$

and set

```
D = diag(diag(A))
E = A − D
```

(a) To see how the Gerschgorin circles and eigenvalues of $A(t)$ change as t goes from 0 to 1 set

```
t = [0, 0.25, 0.50, 0.75, 0.85, 0.95, 1]
color = 'ymcrgbw'
```

For each of the seven values of t, plot the Gerschgorin circles and eigenvalues of $A(t)$ using the seven colors specified by the string **color**. Hold each of the seven plots on the same figure. The plots can be generated using the following commands:

```
for j = 1:7
    B = D + t(j) * E;
    gersch(B,1,color(j))
    hold on
    pause(2)
end
```

(b) One of the Gerschgorin circles of $A = A(1)$ does not contain any eigenvalues of A. Determine the first value of j for which the eigenvalue of $A(t(j))$ no longer remains within the jth circle. Also determine the first value of j for which the eigenvalues of $A(t(j))$ are no longer real.

25. Provide an argument based on your observations from problems 23 and
24 which proves that if the first k Gerschgorin circles of $A = A(1)$ are
disjoint from the others, then for any t, $0 \le t \le 1$, the first k Gerschgorin
circles of $A(t)$ will always contain exactly k eigenvalues of $A(t)$.

Chapter 8

Singular Value Decomposition

8.1   Singular Value Decomposition Exercises

Any $m \times n$ matrix A can be factored into a product of the form USV^{T}, where U and V are orthogonal matrices and S is an $m \times n$ diagonal matrix whose diagonal entries are all real and satisfy

$$s_1 \geq s_2 \geq \cdots \geq s_k \geq 0 \qquad k = \min(m, n)$$

The factorization USV^{T} is called the *singular value decomposition* (SVD) of A. The diagonal entries $s_1, s_2, \ldots, s_k$ are the *singular values* of A. The column vectors of V are called the *right singular vectors* of A, and the column vectors of U are called the *left singular vectors* of A. The singular value decomposition is a useful tool for solving many applied problems.

1. **Singular Values and Numerical Rank.**
   In this exercise we explore the relation between the singular values of a matrix and its rank.

   (a) Use the ATLAST command

   $$A = \mathsf{randint(8,5,9,3)}$$

161

to generate an 8×5 random integer matrix of rank 3 with entries between -9 and 9. Compute the singular values of A. Do you notice any relationship between the singular values and the rank of a matrix?

(b) For different values of m, n, r with $r \leq \min(m, n)$, use the **randint** command to generate $m \times n$ matrices of rank r; in each case compute the singular values of the matrix. Make a conjecture about the relationship between singular values and rank.

(c) If A has singular value decomposition USV^{T}, how does the rank of A compare to the rank of S? (If a matrix is multiplied by a nonsingular matrix, what is the effect on the rank?) Prove your conjecture from part (b).

(d) The $n \times n$ Hilbert matrix H is a matrix whose (i, j) entry is given by

$$h_{ij} = \frac{1}{i + j - 1}$$

It can be generated by the MATLAB command $H = \mathsf{hilb(n)}$.

   i. For $n = 10$, 12, 14, and 16, generate the $n \times n$ Hilbert matrix. In each case compute the inverse and rank of the matrix using MATLAB's **inv** and **rank** commands. Which of these Hilbert matrices appear to be singular?

   ii. The MATLAB command **rank** determines the *numerical rank* of a matrix rather than its exact rank. The numerical rank of a matrix is determined by its singular values. If k of the singular values of an $n \times n$ matrix are very close to 0, then for finite precision numerical computations the matrix will behave just as if those singular values were 0. For computational purposes, the matrix is indistinguishable from a rank $n - k$ matrix. Consequently, we say that its *numerical rank* is $n - k$. (For a more precise definition of numerical rank, type **help rank**.)
   In actuality, all of the Hilbert matrices are nonsingular, however, the larger n is, the closer the matrix will be to one of lower rank. Compute the singular values of the 12×12 Hilbert matrix H and display the results using **format long**. If all displayed digits of a singular value are 0, then the computed singular value is 0 to machine precision even though it would be nonzero if calculated in exact arithmetic. How many of the computed singular values of H were nonzero? Use the **rank** command to determine the numerical rank of H. Repeat this experiment with the 16×16 Hilbert matrix.

What relationship do you observe between the computed singular values and the numerical rank?

2. **Rank Deficient Least-Squares Problems.**
The least-squares problem of finding a vector x to minimize $\|b - Ax\|$ will have a unique solution if A is an $m \times n$ matrix of rank n. If the rank of A is less than n, the least-squares problem will have infinitely many solutions. Often it is desirable to find the solution x with smallest norm. In this exercise we will learn how to use the singular value decomposition to find the solution with smallest norm. Generate a random 8×6 matrix of rank 4 and a random vector in $\mathbf{R}^8$ using the commands

```
A = randint(8,6,5,4)
b = randint(8,1)
```

Compute the singular value decomposition of A. using the command

```
[U,S,V] = svd(A)
```

Compute a random vector x in $\mathbf{R}^6$ to serve as a possible solution of $Ax = b$, and the compute $y = V^T x$ and $c = U^T b$.

(a) Use the MATLAB command **norm** to verify that

$$\|y\| = \|x\| \quad \text{and} \quad \|c - Sy\| = \|b - Ax\|$$

(b) Show that in general if A is an $m \times n$ matrix with singular value decomposition $A = USV^T$, and x and b are arbitrary vectors in $\mathbf{R}^n$ and $\mathbf{R}^m$, respectively, y and c are defined by $y = V^T x$ and $c = U^T b$ then

$$\|y\| = \|x\| \quad \text{and} \quad \|c - Sy\| = \|b - Ax\|$$

(Hint: See Exercise 1(d) in Chapter 6.)

(c) If $A = USV^T$ is an $m \times n$ matrix of rank r, then the last $m - r$ rows of S will consist entirely of 0's. The vector $c - Sy$ must then be of the form

$$c - Sy = \begin{pmatrix} c_1 - s_1 y_1 \\ c_2 - s_2 y_2 \\ \vdots \\ c_r - s_r y_r \\ c_{r+1} \\ c_{r+2} \\ \vdots \\ c_m \end{pmatrix}$$

The value of $\|c - Sy\|$ is clearly independent of $y_{r+1}, \ldots, y_n$. How should $y_1, y_2, \ldots, y_r$ be chosen in order to minimize $\|c - Sy\|$?
Hint:

$$\|c - Sy\|^2 = \sum_{i=1}^{r} (c_i - s_i y_i)^2 + \sum_{i=r+1}^{m} c_i^2$$

Although the values of $y_{r+1}, \ldots, y_n$ can be arbitrary, the choice of $y_{r+1} = \cdots = y_n = 0$ will correspond to the solution vector x with the smallest norm. (Recall that $\|x\| = \|y\|$.) Let S^+ be the matrix formed by transposing S and replacing each of its nonzero diagonal entries by their reciprocals. Show that the y vector we have constructed is equal to S^+c. The matrix S^+ is called the *pseudoinverse* of S. Show that the solution x to the least-squares problem that has minimum norm is given by

$$x = V S^+ U^T b$$

(d) The MATLAB command pinv(C) can be used to compute the pseudoinverse of any matrix C. Use the pinv command to compute the pseudoinverses, S^+ and A^+, of S and A. Compute also the product $V S^+ U^T$ and compare it to A^+. They should be equal. (In fact, the MATLAB routine pinv computes the pseudoinverse of a nondiagonal matrix A by computing its singular value decomposition $U S V^T$ and setting $A^+ = V S^+ U^T$.) Use the pseudoinverse to compute the solution x to the least-squares problem which has minimum norm.

(e) Computer another solution to the least-squares problem by setting

z = A\b

Compute $\|x\|$ and $\|z\|$ and compare the two quantities. Also compare $\|b - Ax\|$ and $\|b - Az\|$.

3. Pseudoinverses and Linear Systems

For each of the following systems $A\mathbf{x} = \mathbf{b}$, compute the pseudoinverse of
the coefficient matrix and use it to find the smallest least-squares solution
$\mathbf{x}$ to the system. If the system is consistent, the solution calculated using
the pseudoinverse will be the smallest exact solution. Also compute a
least-squares solution $\mathbf{z}$ for each system by setting $\mathbf{z} = A\backslash\mathbf{b}$. Compute
$A\mathbf{x}$ and $A\mathbf{z}$ and verify that these vectors are equal even when $\mathbf{x} \neq \mathbf{z}$. In
each case, test to see whether or not the calculated solutions are exact
solutions by computing the residual $\mathbf{b} - A\mathbf{x}$. (Note: it is not necessary
to compute the residual for the solution $\mathbf{z}$ since it should be equal to the
residual for $\mathbf{x}$.) In general what conditions must A satisfy to guarantee
$\mathbf{x} = \mathbf{z}$. (When does $A\mathbf{x} = A\mathbf{z}$ imply $\mathbf{x} = \mathbf{z}$?)

(a)

$$-x_1 + 0x_2 + 0.5x_3 + x_4 = 1$$
$$x_1 + x_2 + x_3 + x_4 = 2$$

(b)

$$x_1 + x_2 + x_3 = 1$$
$$2x_1 + x_2 - x_3 = 2$$
$$x_1 + 3x_2 + 2x_3 = 3$$
$$x_1 - 2x_2 + 4x_3 = 4$$

(c)

$$x_1 + 3x_2 - 2x_3 = 3$$
$$2x_1 + 6x_2 - 4x_3 = 2$$
$$-3x_1 - 9x_2 + 6x_3 = 1$$
$$-x_1 - 3x_2 + 2x_3 = 0$$

(d)

$$x_1 - 4x_2 + x_3 + 3x_4 = 1$$
$$x_1 - 3x_2 + x_3 + 2x_4 = 2$$
$$-2x_1 + 3x_2 - 2x_3 - 1x_4 = 1$$
$$2x_1 - 3x_2 + 2x_3 + x_4 = 2$$
$$3x_1 - 6x_2 + 3x_3 + 3x_4 = 1$$

4. Algebraic Properties of Pseudoinverses

(a) If A is a nonsingular $n \times n$ matrix, then $A^{-1}AA^{-1} = A^{-1}$. Does the same property hold for pseudoinverses? To see whether or not it does, set

```
A = randint(6,4,8,2)
B = pinv(A)
```

and compute BAB. Is the result equal to B? Repeat this experiment for A = randint(m,n,k,r) with a variety of values for m,n,k, and r. Does $BAB = B$ in each case?

(b) If A is a nonsingular $n \times n$ matrix, then $AA^{-1}A = A$. Does the same property hold for pseudoinverses? Test to see whether this property holds in general by generating a number of random integer matrices. For each matrix A, set B = pinv(A) and then compute ABA.

(c) If A is a nonsingular $n \times n$ matrix, then $AA^{-1} = I$. Does the same property hold for pseudoinverses? To see whether or not it does, try a number of examples. Also, if B is the pseudoinverse of A, how do AB and $(AB)^{\mathrm{T}}$ compare?

(d) If A is a nonsingular $n \times n$ matrix, then $A^{-1}A = I$. Does the same property hold for pseudoinverses? To see whether or not it does, try a number of examples. Also, if B is the pseudoinverse of A, how do BA and $(BA)^{\mathrm{T}}$ compare?

(e) Prove all of the identities you discovered in parts (a) through (d).

8.2 Singular Value Decomposition Projects

1. Visualizing the Singular Value Decomposition

In the simple case that A is a 2×2 matrix it is not difficult to visualize the singular values and singular vectors of A. The matrix A can be thought of as a linear transformation T mapping $\mathbf{R}^2$ into $\mathbf{R}^2$; i.e., $T(\mathbf{x}) = A\mathbf{x}$ for all $\mathbf{x}$ in $\mathbf{R}^2$. One can visualize the effect of T by applying T to all of the points of a figure in the plane and then graphing the transformed figure. The command eigshow(A) can be used to visualize the image of the unit circle under T. As an example, set A = [1 1; 0 2] and then enter the command eigshow(A). To see the image of the unit circle under the linear transformation defined by the matrix A, hold down the left mouse button and move the vector $\mathbf{x}$ around the unit circle by dragging the mouse in a counterclockwise circle. For each vector $\mathbf{x}$, the image $A\mathbf{x}$ is plotted. The image points should form an ellipse.

(a) Use the eigshow command for each of the matrices listed below. In each case try to visualize where the eigenvectors of A are. Is the image always an ellipse? For each ellipse, determine if the eigenvectors of A are in the direction of the axes of the ellipse. If so, are the eigenvalues the lengths of the axes of the ellipse?

     i. A = [2 3; 3 2]
    ii. A = [1 1; 0 1]
   iii. A = [1 6; 3 2]
   iv. A = [2 0; 0 1]
    v. A = [2 1; 4 2]

(b) Let A be a 2×2 matrix with singular value decomposition USV^{T} where U, S, V are of the form

$$U = [\mathbf{u}_1, \ \mathbf{u}_2] \qquad S = \begin{pmatrix} s_1 & 0 \\ 0 & s_2 \end{pmatrix} \qquad V = [\mathbf{v}_1, \ \mathbf{v}_2]$$

In part (c) we will relate the singular value decomposition to the image of the linear transformation defined by A. Before doing this, it is helpful to establish some results about the singular value decomposition. Prove each of the following results or explain why they are true.

    i. $AV = US$.

    ii. $A\mathbf{v}_1 = s_1\mathbf{u}_1$ and $A\mathbf{v}_2 = s_2\mathbf{u}_2$.

    iii. $\mathbf{v}_1$ and $\mathbf{v}_2$ are orthogonal unit vectors and the image vectors $A\mathbf{v}_1$ and $A\mathbf{v}_2$ are also orthogonal.

    iv. $\|A\mathbf{v}_1\| = s_1$ and $\|A\mathbf{v}_2\| = s_2$

(c) We saw in part (b) that the right singular vectors $\mathbf{v}_1$ and $\mathbf{v}_2$ of A are orthonormal vectors and their images are orthogonal. To find $\mathbf{v}_1$ and $\mathbf{v}_2$, we can start with the graph of any pair of orthonormal vectors $\mathbf{x}$ and $\mathbf{y}$ and look at the images $A\mathbf{x}$ and $A\mathbf{y}$. If the images are not orthogonal, we can rotate $\mathbf{x}$ and $\mathbf{y}$ simultaneously until we find a pair of rotated vectors whose images are orthogonal. This can be done clicking on the *eig/(svd)* button in the **eigshow** graphics window. The svd mode works just like the **eigshow** mode, except you use the mouse to rotate a pair of orthonormal vectors rather than a single vector. The images of the two vectors are plotted and the angle between the images (measured in degrees) is displayed. Use the svd mode of **eigshow** to determine visually the right singular vectors for each of the matrices in part (a). Can you estimate visually the corresponding singular values? Explain. How do the singular values and singular vectors of A relate to the axes of the ellipses?

(d) If A has singular value decomposition USV^{T}, determine the singular value decomposition of A^{T}. For each matrix A in part (a), set $\mathsf{A} = \mathsf{A}'$ and run **svdshow**. How do the axes of the ellipses correspond to the singular values and singular vectors of A^{T}? How do they correspond to the singular values and singular vectors of the original matrix A?

2. The SVD and Digital Image Processing

To digitize an image means to break the image into a grid of many tiny rectangles and assign a number to each rectangle. The numbers represent colors that are associated with each rectangle in the image. Therefore a fine grid with many rectangles will give a better representation of the image than a coarse grid with just a few rectangles, but the fine grid will require many more numbers. These numbers can be stored in a matrix of the same size as the grid. Thus an image with a 1000×1000 grid would yield a 1000×1000 matrix consisting of one million entries. If we want to store this image on a computer or send it across the country, we could store or send the matrix and use the numbers later to reconstruct the original image. The problem, of course, is that if we have many such images, the space involved in storing such large matrices, or the time required to transmit so much data, could become excessive.

What is needed is a way to compress the data so that the original image, or a reasonable approximation of the image, could be reconstructed with much less data. The singular value decomposition provides a way to do this. Suppose $A = USV^T$ is the singular value decomposition of the $m \times n$ matrix A, where U and V are orthogonal matrices and S is an $m \times n$ diagonal matrix with singular values along the diagonal in decreasing numerical order. Let $k = \min(m, n)$ and let s_j $(j = 1, \ldots, k)$ denote the jth singular value along the diagonal of S. If $\mathbf{u}_j$ and $\mathbf{v}_j$ represent the jth column vectors of U and V, respectively, then A can be written as an outer product expansion

$$A = s_1 \mathbf{u}_1 \mathbf{v}_1^T + s_2 \mathbf{u}_2 \mathbf{v}_2^T + \cdots + s_k \mathbf{u}_k \mathbf{v}_k^T$$

By truncating this expansion, we can obtain the best approximations to A by matrices of lower rank. Because we are using the largest singular values first, we hope that most of the information contained in A will be reproduced using relatively few terms of the expansion. We expect then that a matrix of the form

$$A_r = s_1 \mathbf{u}_1 \mathbf{v}_1^T + s_2 \mathbf{u}_2 \mathbf{v}_2^T + \cdots + s_r \mathbf{u}_r \mathbf{v}_r^T$$

will adequately represent the original image given by A even if r is much smaller than k. If $s_r > 0$, then A_r is a rank r approximation to A.

In this project we will first work with small matrices and geometric images to help understand how the SVD can be used to create approximations to an image. The MATLAB command **image(C)** displays the matrix C as an image. Each entry of C is interpreted as a color of a rectangular patch in the image. The elements of C are used as indices into the current *colormap* to determine the color that should be displayed. MATLAB's default colormap has entries for 64 shades of color. The first entry of this colormap corresponds to the color red, so any entry of C that is 1 would correspond to a red rectangle in the image of C. Some monitors may not be able to display all 64 shades of color. To see how your monitor works, create a matrix C with the integers 1 to 64 and then display the image represented by this matrix. This can be done with the commands

```
C = 1:64;  image(C)
```

To read more about these issues, use the MATLAB help facility for the key words **color** and **colormap**.

(a) Create a diagonal matrix by setting A = 26*eye(3). The diagonal
entries of A all equal 26 which represents one of the 64 colors in the
color map. What is the rank of A? Type the MATLAB command

[U,S,V] = svd(A)

This command generates the three factors in the singular value decomposition of A. Now enter the command

svdimage(A,U,S,V)

This will generate two images. The image on the left is the one represented by the matrix A. The image on the right is the representation
of the rank one approximation to A using only the largest singular
value of A. Thus the image on the right corresponds to the matrix
$A_1 = s_1 \mathbf{u}_1 \mathbf{v}_1^T$. While the mouse is positioned in the figure window,
click the left mouse button. The image on the right will now be the
rank 2 approximation

$$A_2 = s_1 \mathbf{u}_1 \mathbf{v}_1^T + s_2 \mathbf{u}_2 \mathbf{v}_2^T$$

Click the mouse button once again to see the image using all three
singular values. This should of course be the same as the original
matrix since A has rank 3.

We want to understand what is happening here. Look at the matrix
S. What are the three singular values of A? What are the matrices U
and V in this example? Compute the product $A_1 = s_1 \mathbf{u}_1 \mathbf{v}_1^T$ by setting

A1 = s(1,1)*u(:,1)*v(:,1)'

Do you see why the image of the first approximation only had one box
in the upper left hand corner? Use the command image(A1) to view
the image of A_1. Now compute the product $s_2 \mathbf{u}_2 \mathbf{v}_2^T$ and add this to
the first approximation. What does the image for this matrix look
like?

Now let's repeat this experiment with a larger matrix. Enter the
commands

A = 26*eye(10);
[U,S,V] = svd(A);

(Be sure to use the semi-colon to suppress output.)

Run **svdimage(A,U,S,V)** and describe how the successive approximations recreate the original image. In view of your calculations for the 3×3 example, explain what is happening for this 10×10 example.

(b) Create a diagonal matrix using the command

$$A = \mathsf{diag(randint(1,20,20)+30)};$$

The diagonal entries of A are random integers between 10 and 50. Compute the SVD for this matrix and display the images for the approximations to A computed from U, S, and V. Describe the manner in which the approximations are converging to the original image. What is different in this case from part (a)? What type of matrices are U and V? Can you explain the order in which the rectangles are being drawn in each successive approximation? To help understand this, use the commands

```
C = 10:50;
image(C)
```

to generate an image of the colormap from 10 to 50. Move this figure window to the bottom of the screen. Rerun **svdimage** for the matrix A in another window at the top of the screen. Remember that the singular values are used in decreasing numerical order to form the approximation images.

(c) To generate an image of a letter X, set $A = 26*\mathsf{Xmatrix(10)}$ and compute its singular value decomposition. Use the **svdimage** command to examine the first five approximate images. What happens by the fifth approximation? What is the rank of A? Why is A not of full rank? What are the singular values of A? How does the rank explain what happened with the approximations?

(d) Here are three more images to try. For each, experiment with the SVD approximations to see how well they convey the information in the original picture. Don't forget to use the semi-colon to suppress the printing of the large matrices!

 i. Create a band matrix using the commands

$$S = \text{diag}(26*\text{ones}(29,1),1);$$
$$A = S + S' + 26*\text{eye}(30);$$

ii. Using the matrix A from part (i), create a "thick" letter X matrix by setting

$$X = A + \text{fliplr}(A);$$

iii. Create a random 20×20 image using the command

$$A = \text{randint}(20)+10;$$

(e) The command **niceimag** generates a 12×70 mystery matrix and uses **svdimage** to display the image corresponding to its rank one svd approximation. Discover the message contained in the mystery matrix by clicking the mouse to increase the rank of the approximation. Why should you be able to determine the image completely with at most eleven clicks? How many clicks are actually necessary? What is the rank of the original matrix? Is it possible to determine the number of linearly independent rows in the original matrix by visually examining the final image? Explain.

(f) There are two M-files you can use to create your own simple images. The command

$$A = \text{makeimag}(n,\text{imag},\text{back});$$

lets you create an $n \times n$ image matrix A using **imag** and **back** as the image and background colors, respectively. After entering the command, place the mouse pointer at any point in the graph and click the left mouse button to toggle a square between the image and background colors. Once you have created the matrix A, the command

$$A = \text{editimag}(A,\text{imag},\text{back})$$

allows you to edit it further. Use these two M-files to create an image of your choice. For example, your image could be a stick figure, a polygon, your initials, the grade you hope to get in the course. Compute the singular value decomposition of your image matrix and use **svdimage** to investigate the SVD approximations to the image.

Realistic images require very large image matrices. Each entry of the image matrix will correspond to a small dot on the screen. Unfortunately, the student version of MATLAB restricts the size of the matrices we can use. However, if the professional version of MATLAB is available, it is possible to study images corresponding to larger matrices. The remainder of this project requires the professional version of MATLAB.

In examining the SVD images of the letter X matrix in part (c), you might have noticed that the rank three approximation resembles the letter X even though some of the blocks on the diagonal and antidiagonal had not yet been filled in. The positions of the missing blocks are convenient because often our senses will recognize the pattern for us and fill in what is missing. To test this, create a letter X matrix by setting A=26*Xmatrix(100). Compute the singular value decomposition of A and then generate the image obtained from the first 25 terms of the outer product expansion using the command

svdimage(A,U,S,V,25)

Does your eye see the "same" image as the original X? The original image requires a matrix with 10,000 entries. How many entries are required to compute your approximation image? If, for example, you had used only 5 terms of the outer product expansion, you would have used 10 vectors, each with 100 entries, and 5 singular values for a total of 1005 numbers instead of the original 10,000 numbers.

(g) We would now like to get a sense of how SVD approximations could be used with some real images. Enter the command **load penny**. This will create a 128×128 matrix P. To get a good image, modify P by rescaling its entries so that all of them are between 1 and 64; i.e., set

P=ceil(64*P/max(max(P)));

Compute the SVD of P. (Be sure to include the semi-colon to suppress output.) Be patient–P is a large matrix, so it could take a while to compute its SVD. Use the command

svdimage(P,U,S,V,1,copper)

to investigate successive approximations to the image. Stop when you feel you have a reasonable approximation to the original image. What is the rank of your approximation? How many entries are necessary to construct this approximation? What percentage is this of the total number of entries used for the original image?

(h) Enter the command **load gatlin2**. This will create a 176×260 matrix X. The image comes with its own colormap map matrix, appropriately called **map**. The commands

    image(X), colormap(map)

will generate the image of a black and white photo of three pioneers of numerical analysis. They are (from left to right) James Wilkinson, Wallace Givens, and George Forsythe. Compute the SVD of X. (Be patient–this could take a while.) Use the command

    svdimage(X,U,S,V,1,map)

to investigate successive approximations to the image. Stop when you feel you have a reasonable approximation to the original image. What is the rank of your approximation? How many entries are necessary to construct this approximation? What percentage is this of the total number of entries used for the original image?

One can attempt to colorize the photo using the colormap **hot**. Use the command

    svdimage(X,U,S,V,1,hot)

to generate SVD approximations to the colorized photo. Answer the same questions as before for the colorized approximations.

3. Circles in Space

If A is an $m \times n$ matrix of rank r with singular value decomposition $A = USV^{\mathrm{T}}$, then the first r columns of U form an orthonormal basis for the column space of A. The remaining $m-r$ column vectors of U are each orthogonal to any vector in the column space of A. This observation is the key to the application presented in the following project. The project is based on a paper by Carl Cowen [1].

In Project 1 of Section 6.2 we talked about machines used to measure coordinates along the perimeter of a manufactured part. The measurements are used by quality control engineers to determine whether or not the perimeter is circular. The methods employed in that project assumed that the measured points lie in the xy-plane. However, in many cases the measurements are taken with respect to an xyz coordinate system; that is, three coordinates are specified for each measured point. In these cases, in order to determine if the points lie on a circle, we must first transform the data to the xy-plane.

Suppose we are given a set of points $\{(x_i, y_i, z_i)\}$ that lie on a circle contained in some plane other than the xy-plane. In order to describe this circle, not only do we need to specify the center and radius, but we must also characterize the plane containing the circle. Any plane can be given by an equation of the form

$$a(x - x_0) + b(y - y_0) + c(z - z_0) = 0$$

where a, b, c are the components of a vector normal to the plane, and x_0, y_0, z_0 are the coordinates of a point in the plane. Now if the given set of points lies in the plane, the circle through the points must have its center in the plane as well. So we can use the center of the circle as the point (x_0, y_0, z_0). Once we have determined this center point, the radius r, and a normal vector $[a, b, c]$, we will have described the circle in space.

In order to perform a least-squares analysis, we must first find some affine map that transforms the points to the xy-plane. Once we have found the center and radius of the transformed circle, we can reverse the map to determine the center of the original circle and the plane containing it.

(a) Consider the collection of ten points whose x, y, z coordinates are given by:

x	-6.30	-4.84	-2.97	2.26	4.17	3.67	2.92	1.14	-0.82	-6.41
y	1.69	-1.46	-2.74	-0.77	7.11	8.71	9.98	11.40	11.68	4.48
z	1.11	3.08	5.10	9.70	10.02	9.25	8.29	6.33	4.42	0.47

Enter the x, y, z coordinates of these points in MATLAB as row vectors x, y, and z and set C = [x; y; z].

Not only do these points lie off the xy-plane, they are contained in a plane that likely does not pass through the origin. The first step will be to translate the points to another plane that does pass through the origin. We could do this by subtracting the coordinates of any one of

the given points. However, for the sake of stability, we will determine
the average of all the points (i.e., their *center of mass*) and subtract
its coordinates from each of the given points. To do this, set

average = sum(C′)′/10
D = C − average∗ones(1,10)

The columns of D contain the coordinates of the translated points;
these points should lie on a circle of the same radius as the original
circle. Moreover, the translated points should (more or less) lie on a
plane that passes through the origin.

This latter plane, however, is still not the xy-plane. In fact, it has the
same normal vector as the original plane. In order to translate this
plane to the xy-plane, compute the singular value decomposition of D
by setting

[U,S,V] = svd(D)

The first two columns of U give an orthonormal basis of the plane
containing the points in D. The third column of U is orthogonal to
the plane and thus provides the normal vector. To keep track of this
vector, set **N = U(:,3)**.

If all the points were exactly planar, then the matrix D would have
rank 2 and its third singular value s_3 would be exactly 0. When s_3 is
nonzero, it can be used as a measure of how close D is to a matrix of
rank 2. Thus, if s_3 is small, the given points are nearly planar.

We can transform the matrix D using the orthogonal matrix U^T. Since
$D = USV^T$ it follows that $U^TD = SV^T$. So to determine the translated points set

R = S∗V′

The columns of R contain the translated points which (more or less)
lie in the xy-plane. The third row of R is $s(3)\mathbf{v}_3^T$. If s_3 is small, then
the entries of the third row should all be close to 0.

Extract the x and y coordinates from the matrix R by setting

x1 = R(1,:)′;   y1 = R(2,:)′

These points in the xy-plane lie on a circle of the same radius as the original circle. We cannot assume that the center of this transformed circle is necessarily at the origin. We can determine the center **c1** and radius **r** of the circle by performing a least-squares analysis using the method given in Project 1 of Section 6.2. To do this, set

A = [2*x1, 2*y1, ones(size(x1))]
b = x1.^2 + y1.^2
c1 = A\b
r = sqrt(c1(3) + c1(1)^2 + c1(2)^2)

At this point it may be helpful to check visually how well the transformed points lie on a circle. Plot the data points given by the vectors **x1** and **y1** and the graph of the least-squares circle, all on the same axis system.

To get the center of the original circle, we must translate back to our original coordinate system. To do this set

c = c1(1)*U(:,1) + c(2)*U(:,2) + average

Measure the total deviation by setting

s = sqrt((x − c(1)).^2 + (y − c(2)).^2+(z − c(3)).^2);
e = s − r
d = norm(e)

Use the coordinates of the center of the original circle and the components of the normal vector to write the equation of the plane.

(b) Repeat part (a) for each of the following sets of points.

(i)

x	−13.10	−4.55	0.93	3.03	2.98	−2.07	−14.41	−16.79	−23.73
y	−20.18	−21.16	−21.06	−20.65	−18.12	−16.29	−14.41	−14.33	−15.71
z	−39.86	−39.08	−35.12	−31.80	−19.84	−14.35	−13.28	−14.44	−25.41

(ii)

x	−105.65	−58.31	70.95	109.18	116.33	118.64	73.54	−74.34	−129.79
y	−80.51	−100.02	−70.80	−20.86	1.10	39.81	123.74	150.70	116.52
z	−30.27	−38.76	−25.84	−3.94	5.67	22.62	59.32	70.97	55.95

Appendix A

MATLAB

MATLAB is an interactive program for matrix computations. The original version of MATLAB, short for *matrix laboratory* was developed by Cleve Moler from the Linpack and Eispack software libraries. Over the years MATLAB has undergone a series of expansions and revisions. Today it is the leading software for scientific computations. The professional version of MATLAB is distributed by the Math Works, Inc. of Natick, Massachusetts.

Basic Data Elements

The basic elements that MATLAB uses are matrices. Once the matrices have been entered or generated, the user can quickly perform sophisticated computations with a minimal amount of programming.

Entering matrices in MATLAB is easy. To enter the matrix

$$\begin{pmatrix} 1 & 2 & 3 & 4 \\ 5 & 6 & 7 & 8 \\ 9 & 10 & 11 & 12 \\ 13 & 14 & 15 & 16 \end{pmatrix}$$

one could type

A = [1 2 3 4; 5 6 7 8; 9 10 11 12; 13 14 15 16]

or the matrix could be entered one row at a time:

A = [1   2   3   4
        5   6   7   8
        9   10   11   12
       13   14   15   16]

Row vectors of equally spaced points can be generated using MATLAB's
: operation. The command x = 2:6 generates a row vector with integer
entries going from 2 to 6.

x =

             2 3 4 5 6

It is not necessary to use integers or to have a stepsize of 1. For example,
the command x = 1.2:0.2:2 will produce

x =
          1.2000 1.4000 1.6000 1.8000 2.0000

Submatrices

To refer to a submatrix of A, one must use the : to specify the rows and
columns. For example, the submatrix consisting of the entries in the second
two rows of columns 2 through 4 is given by $A(2:3, 2:4)$. Thus the statement

C = A(2 : 3, 2 : 4)

generates

C =

            6   7   8
           10   11   12

If the colon is used by itself for one of the arguments, either all of the rows or all of the columns of the matrix will be included. For example, $A(:, 2:3)$ represents the submatrix of A consisting of all the elements in the second and third columns and $A(4, :)$ denotes the fourth row vector of A.

Generating Matrices

One can also generate matrices using built-in MATLAB functions. For example, the command

$$B = \text{rand}(4)$$

will generate a 4×4 matrix whose entries are random numbers between 0 and 1. Other functions that can be used to generate matrices are **eye, zeros, ones, magic, hilb, pascal, toeplitz, compan,** and **vander**. To build triangular or diagonal matrices one can use the MATLAB functions **triu, tril,** and **diag**.

The matrix building commands can be used to generate blocks of partitioned matrices. For example, the MATLAB command

$$E = [\text{ eye}(2), \text{ ones}(2,3); \text{ zeros}(2), [1:3; \quad 3:-1:1] \text{ }]$$

will generate the matrix

$$
E =
$$

$$
\begin{array}{ccccc}
1 & 0 & 1 & 1 & 1 \\
0 & 1 & 1 & 1 & 1 \\
0 & 0 & 1 & 2 & 3 \\
0 & 0 & 3 & 2 & 1
\end{array}
$$

Matrix Arithmetic

Matrix arithmetic in MATLAB is straightforward. We can multiply our original matrix A times B simply by typing **A*B**. The sum and difference of A and B are given by **A + B** and **A − B**, respectively. The transpose of A is given by **A'**. If **c** represents a vector in R^4, the solution to the linear system $A\mathbf{x} = \mathbf{c}$ can be computed by setting

$$x = A \backslash c$$

Powers of matrices are easily generated. The matrix A^5 is computed in MATLAB by typing **A^5**. One can also perform operations elementwise by preceding the operand by a period. For example, if **W** = [1 2; 3 4], then **W^2** results in

$$
\text{ans} =
$$

$$
\begin{array}{cc}
7 & 10 \\
15 & 22
\end{array}
$$

while W.^2 will give

$$
\textbf{ans} =
$$

$$
\begin{matrix}
1 & 4 \\
9 & 16
\end{matrix}
$$

MATLAB Functions

To compute the eigenvalues of a square matrix A, one need only type **eig(A)**. The eigenvectors and eigenvalues can be computed by setting

$$
[\textsf{X} \quad \textsf{D}] = \textbf{eig(A)}
$$

Similarly, one can compute the determinant, inverse, condition number, norm, and rank of a matrix with simple one-word commands. Matrix factorizations such as the LU, QR, Cholesky, Schur decomposition, and singular value decomposition can be computed with a single command. For example, the command

$$
[\textsf{Q} \quad \textsf{R}] = \textbf{qr(A)}
$$

will produce an orthogonal (or unitary) matrix Q and an upper triangular matrix R, with the same dimensions as A, such that $A = QR$.

Programming Features

MATLAB has all the flow control structures that you would expect in a high-level language including **for** loops, **while** loops, and if statements. This allows the user to write his or her own MATLAB programs and to create additional MATLAB functions. It should be noted that MATLAB prints out automatically the result of each command unless the command line ends in a semicolon. When using loops we recommend ending each command with a semicolon to avoid printing all of the results of the intermediate computations.

Relational and Logical Operators

MATLAB has six relational operators which are used for comparisons of scalars or elementwise comparisons of arrays. These operators are

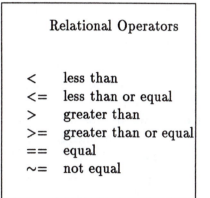

Given two $m \times n$ matrices A and B, the command

$$C = A < B$$

will generate an $m \times n$ matrix consisting of zeros and ones. The (i, j) entry will be equal to 1 if and only if $a_{ij} < b_{ij}$. For example, suppose that

$$A = \begin{pmatrix} -2 & 0 & 3 \\ 4 & 2 & -5 \\ -1 & -3 & 2 \end{pmatrix}$$

The command $A >= 0$ will generate

$$\mathsf{ans} =$$

$$\begin{matrix} 0 & 1 & 1 \\ 1 & 1 & 0 \\ 0 & 0 & 1 \end{matrix}$$

There are three logical operators as shown in the following table.

Logical Operators	
&	AND
|	OR
~	NOT

These logical operators regard any nonzero scalar as corresponding to TRUE and 0 as corresponding to FALSE. The operator & corresponds to the logical

AND. If a and b are scalars, the expression a&b will equal 1 if a and b are both nonzero (TRUE) and 0 otherwise. The operator $|$ corresponds to the logical OR. The expression a|b will have the value 0 if both a and b are 0 and otherwise it will be equal to 1. The operator $\tilde{}$ corresponds to the logical NOT. For a scalar a it takes on the value 1 (TRUE) if $a = 0$ (FALSE) and the value 0 (FALSE) if $a \neq 0$ (TRUE).

For matrices these operators are applied elementwise. Thus if A and B are $m \times n$ matrices, then A&B is a matrix of zeros and ones whose ijth entry is $a(i,j)\&b(i,j)$. For example, if

$$A = \begin{pmatrix} 1 & 0 & 1 \\ 0 & 1 & 1 \\ 0 & 0 & 1 \end{pmatrix} \quad \text{and} \quad B = \begin{pmatrix} -1 & 2 & 0 \\ 1 & 0 & 3 \\ 0 & 1 & 2 \end{pmatrix}$$

then

$$A\&B = \begin{pmatrix} 1 & 0 & 0 \\ 0 & 0 & 1 \\ 0 & 0 & 1 \end{pmatrix}, \quad A|B = \begin{pmatrix} 1 & 1 & 1 \\ 1 & 1 & 1 \\ 0 & 1 & 1 \end{pmatrix}, \quad \tilde{}A = \begin{pmatrix} 0 & 1 & 0 \\ 1 & 0 & 0 \\ 1 & 1 & 0 \end{pmatrix}$$

The relational and logical operators are often used with if statements.

Columnwise Array Operators

MATLAB has a number of functions which when applied to either a row or column vector **x** return a single number. For example, the command max(**x**) will compute the maximum entry of **x**, and command sum(**x**) will return the value of the sum of the entries of **x**. Other functions of this form are min, prod, mean, all, and any. When used with a matrix argument these functions are applied to each column vector and the results are returned as a row vector. For example, if

$$A = \begin{pmatrix} -3 & 2 & 5 & 4 \\ 1 & 3 & 8 & 0 \\ -6 & 3 & 1 & 3 \end{pmatrix}$$

then

$$\text{min}(A) = [-6, 2, 1, 0]$$
$$\text{max}(A) = [1, 3, 8, 4]$$
$$\text{sum}(A) = [-8, 8, 14, 7]$$
$$\text{prod}(A) = [18, 18, 40, 0]$$

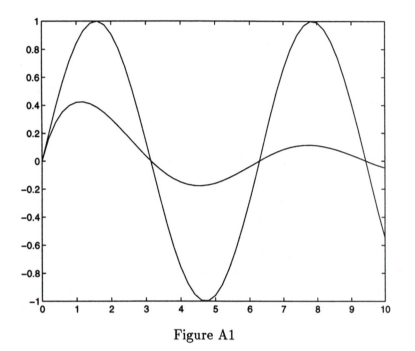

Figure A1

Graphics

If **x** and **y** are vectors of the same length, the command **plot(x,y)** will produce
a plot of all the (x_i, y_i) pairs and each point will be connected to the next
by a line segment. If the x-coordinates are taken close enough together, the
graph should resemble a smooth curve. The command **plot(x,y,′x′)** will plot
the ordered pairs with x's but will not connect the points.

For example, to plot the function $f(x) = \dfrac{\sin x}{x + 1}$ on the interval $[0, 10]$ set

$$\mathsf{x} = 0 : 0.2 : 10 \quad \text{and} \quad \mathsf{y} = \sin(x)./(x + 1)$$

The command **plot(x,y)** will generate the graph of the function. To compare
the graph to that of $\sin x$ one could set $\mathsf{z} = \sin(x)$ and use the command

$$\mathsf{plot(x, y, x, z)}$$

to plot both curves at the same time as in Figure A1.

It is also possible to do more sophisticated type of plots in MATLAB,
including polar coordinates, three-dimensional surfaces, and contour plots.

Help Facility

MATLAB includes a HELP facility that lists and describes all of MATLAB's functions, operations, and commands. To obtain information on any of the MATLAB commands, one need only type **help** followed by the name of the command.

Conclusions

MATLAB is a powerful tool for matrix computations which is also user friendly. The fundamentals can be mastered easily, and consequently, students are able to begin numerical experiments with only a minimal amount of preparation. Indeed, the material in this appendix together with the on-line help facility should be enough to get one started.

While this appendix summarizes the features of MATLAB that are most relevant to an undergraduate course in linear algebra, there are many other advanced capabilities that have not been discussed. For details on these features the reader should consult the *The Student Edition of MATLAB* [2].

Appendix B

Special Matrices

This appendix describes the special matrices that appear in the exercises. By special matrices we mean matrices that have some special structure.

A. MATLAB Special Matrices

The following special matrices can be generated using standard MATLAB commands.

1. **Name:** The *zero* matrix, $O = O_{mn}$

   Description: The $m \times n$ matrix of all 0's.

   MATLAB construction: zeros(m,n), or zeros(n) for a square matrix

2. **Name:** The *identity* matrix, $I = I_n$

   Description: The $n \times n$ matrix of 1's on the main diagonal and 0's elsewhere.

   MATLAB construction: eye(n)

3. **Name:** The *ones* matrix, $J = J_{mn}$

   Description: The $m \times n$ matrix of all 1's.

   MATLAB construction: ones(m,n), or ones(n) for a square matrix

4. **Name:** The *Hilbert* matrix

   Description: The $n \times n$ matrix whose (i, j) entry is $\frac{1}{i+j-1}$.

   MATLAB construction: hilb(n)

187

5. **Name**: The *Hadamard* matrix

   Description: An $n \times n$ matrix whose entries are all 1 and -1 with the property that $H^T H = nI$. If $n > 2$, then it must be divisible by 4.

   MATLAB construction: hadamard(n)

6. **Name**: Toeplitz matrices

   Description: Matrices whose entries are constant along any diagonal. A symmetric Toeplitz matrix is completely determined by its first column and a nonsymmetric Toeplitz matrix is completely determined by its first column and first row.

   MATLAB construction: toeplitz(c) or toeplitz(c,r)

7. **Name**: The Vandermonde matrix

   Description: The $n \times n$ matrix determined from a vector **x** by setting its jth column equal to $[x_1^{n-j}, x_2^{n-j}, \ldots, x_n^{n-j}]^T$, for $j = 1, \ldots, n$

   MATLAB construction: vander(x)

MATLAB includes M-files for generating other interesting special matrices such as *magic squares* and *Pascal matrices*. For a complete listing of the MATLAB special matrices on your system type **help specmat** at the MATLAB prompt.

B. ATLAST Special Matrices

The following special matrices can be generated by ATLAST commands. (For more details on these commands see Appendix C.)

8. **Name**: The *grid* matrix

   Description: The $n \times n$ matrix in which every row is $[\, 1\ 2\ \cdots\ n\,]$.

   MATLAB construction: ones(n,1)*[1:n], or gridmat(n)

9. **Name**: The *maximum* matrix

   Description: The $n \times n$ matrix whose (i, j) entry is the maximum of i and j.

   MATLAB construction: maxmat(n)

10. **Name**: The *minimum* matrix

    Description: The $n \times n$ matrix whose (i, j) entry is the minimum of i and j.

    MATLAB construction: minmat(n)

11. Name: The *checkerboard matrix*

Description: The $n \times n$ matrix whose entries are alternately 1 and 0, with 1 in the upper-left corner.

MATLAB construction: checker(n)

12. Name: The *anticheckerboard* matrix

Description: The $n \times n$ matrix whose entries are alternately 1 and 0, with 0 in the upper-left corner.

MATLAB construction: achecker(n)

13. Name: The *sign* matrix

Description: The $n \times n$ matrix whose entries are alternately 1 and -1, with 1 in the upper-left corner.

MATLAB construction: signmat(n)

14. Name: The *backwards identity* matrix

Description: The $n \times n$ matrix of 1's on the antidiagonal and 0's elsewhere.

MATLAB construction: fliplr(eye(n)) or backiden(n)

15. Name: The *Jordan 0–block* matrix

Description: The $n \times n$ matrix of 1's on the diagonal just above the main diagonal, and 0's elsewhere.

MATLAB construction: diag(ones(n-1,1),1), or jordan0(n)

16. Name: The *Jordan 1–block* matrix

Description: The $n \times n$ matrix of 1's on the main diagonal and on the diagonal just above it; 0's elsewhere.

MATLAB construction: eye(n) + jordan0(n)

17. Name: The *cyclic* matrix

Description: The $n \times n$ matrix of 1's on the diagonal just above the main diagonal and in the lower-left corner; 0's elsewhere.

MATLAB construction: cyclic(n)

18. Name: The *consecutive integers* matrix

Description: The $n \times n$ matrix whose entries are the consecutive integers 1 to n^2 arranged in rows.

MATLAB construction: reshape([1:n^2],n,n)' or consec(n)

19. Name: The *Hankel consecutive integers* matrix

Description: The $n \times n$ matrix whose jth row is $[j, j+1 \ldots, j+n-1]$ for $j = 1, \ldots, n$.

MATLAB construction: hconsec(n)

20. Name: The *letter H* matrix

Description: The $n \times n$ matrix (n odd) of all 1's in the first and last columns and in the middle row; 0's elsewhere.

MATLAB construction: Hmatrix(n)

21. Name: The *letter L* matrix

Description: The $n \times n$ matrix of 1's in first column and last row; 0's elsewhere.

MATLAB construction: Lmatrix(n)

22. Name: The *letter N* matrix

Description: The $n \times n$ matrix of all 1's in the first and last columns and on the main diagonal; 0's elsewhere.

MATLAB construction: Nmatrix(n)

23. Name: The *letter T* matrix

Description: The $n \times n$ matrix (n odd) of all 1's in the first row and middle column; 0's elsewhere.

MATLAB construction: Tmatrix(n)

24. Name: The *letter X* matrix

Description: The $n \times n$ matrix of all 1's on the main diagonal and on the antidiagonal; 0's elsewhere.

MATLAB construction: eye(n)|fliplr(eye(n)), or Xmatrix(n)

25. Name: The *letter Y* matrix

Description: The $n \times n$ matrix of 1's on the antidiagonal and halfway along the main diagonal from upper left to the middle; 0's elsewhere.

MATLAB construction: Ymatrix(n)

26. Name: The *letter Z* matrix

Description: The $n \times n$ matrix of all 1's in the first and last rows and on the antidiagonal.

MATLAB construction: Zmatrix(n)

Appendix C

ATLAST Commands

These are the MATLAB commands that have been specially written for this book of ATLAST exercises. The code for all these commands can be downloaded from the ATLAST web page. (See page ix of the Preface for details.) However, for those readers who just want to type in a few of the commands, we list here the code, when it is not lengthy.

achecker

Synopsis

A = achecker(n)

Description

achecker(n) is the $n \times n$ matrix whose entries are alternately 1 and 0, with 0 in the upper-left corner.

Example

achecker(4) is

0	1	0	1
1	0	1	0
0	1	0	1
1	0	1	0

Code

```
function A = achecker(n)
G = ones(n,1)*[1:n];
A = rem(G+G',2);
```

Alphabet Matrices

Synopsis

$$
\begin{aligned}
H &= \text{Hmatrix(n)} \\
L &= \text{Lmatrix(n)} \\
N &= \text{Nmatrix(n)} \\
T &= \text{Tmatrix(n)} \\
X &= \text{Xmatrix(n)} \\
Y &= \text{Ymatrix(n)} \\
Z &= \text{Zmatrix(n)}
\end{aligned}
$$

Description

The alphabet matrices are square matrices whose entries are 0's and 1's. The nonzero entries of the matrices are in the pattern of one of the following letters of the alphabet: H, L, N, T, X, Y, Z. The input argument for Hmatrix, Tmatrix, and Ymatrix must be odd.

Example

The commands **Nmatrix(4)** and **Tmatrix(3)** produce the following matrices.

1	0	0	1	1	1	1
1	1	0	1	0	1	0
1	0	1	1	0	1	0
0	0	0	1			

Code

```
function A = Hmatrix(n)
if rem(n,2) == 0
     error('The input argument for Hmatrix must be odd')
end
A = zeros(n);
A(:,1) = ones(n,1);
A(:,n) = ones(n,1);
A(fix((n+1)/2),:) = ones(1,n);
```

```
function A = Lmatrix(n)
A = zeros(n);
A(:,1) = ones(n,1);
A(n,:) = ones(1,n);

function A = Nmatrix(n)
A = eye(n);
A(:,1) = ones(n,1);
A(:,n) = ones(n,1);

function A = Tmatrix(n)
if rem(n,2) == 0
    error('The input argument for Tmatrix must be odd')
end
A = zeros(n);
A(1,:) = ones(1,n);
A(:,(n+1)/2) = ones(n,1);

A = Xmatrix(n)
A = eye(n) | fliplr(eye(n))

function A = Ymatrix(n)
if rem(n,2) == 0
    error('The input argument for Ymatrix must be odd')
end
A = fliplr(eye(n));
A(1:(n-1)/2,:) = A(1:(n-1)/2) | fliplr(A(1:(n-1)/2));

function A = Zmatrix(n)
A = fliplr(eye(n));
A(1,:) = ones(1,n);
A(n,:) = ones(1,n);
```

backiden

Synopsis

B = backiden(n)

Description

backiden(n) is the $n \times n$ matrix of 1's on the antidiagonal and 0's elsewhere.

Example

backiden(4) is

0	0	0	1
0	0	1	0
0	1	0	0
1	0	0	0

Code

```
function B = backiden(n)
B = fliplr(eye(n));
```

checker

Synopsis

C = checker(n)

Description

checker(n) is the $n \times n$ matrix whose entries are alternately 1 and 0, with 1 in the upper-left corner.

Example

checker(4) is

1	0	1	0
0	1	0	1
1	0	1	0
0	1	0	1

Code

```
function A = checker(n)
G = ones(n,1)*[1:n];
A = rem(G+G'+1,2);
```

cogame

Synopsis

cogame

Description

cogame generates a menu for playing the Coordinate Game. From the menu the player can choose any one of four levels of play or can choose the two-person game. For the basic game, two vectors u and v and a target point X are drawn in the plane. The player is prompted to input scalars **a** and **b** so that the vector **au+bv** has its tip at the point X. The resulting vector is plotted each time the player inputs **a** and **b**. In the two-person version, the first player inputs **u**, **v**, and X, and the second player must find **a** and **b**. The two players then reverse roles, and the game is scored based on the number of guesses. The player with the lower score wins.

Utility files

cogame requires the following ATLAST utility files:

clrarrow, coenter, drawvec, levels, tip.

colbasis

Synopsis

C = colbasis(A)

Description

colbasis(A) is a matrix whose columns form a basis for the column space of A. The basis is obtained from the reduced row echelon form of A.

Example

Suppose A is the 3 × 4 matrix

```
    3    -3     1    -3
   -2     2     1     2
   -1     1    -3     1
```

Then **colbasis(A)** is

```
    3     1
   -2     1
   -1    -3
```

Code

```
function C = colbasis(A)
[R,jp] = rref(A);
C = A(:,jp);
```

consec

Synopsis

C = consec(n)

Description

consec(n) is the $n \times n$ matrix whose entries are the consecutive integers from 1 to n^2 arranged in rows.

Example

consec(3) is

1	2	3
4	5	6
7	8	9

Code

```
function A = consec(n)
A = reshape([1:n^2],n,n)';
```

cyclic

Synopsis

C = cyclic(n)

Description

cyclic(n) is the $n \times n$ matrix of 1's on the diagonal just above the main diagonal and in the lower-left corner; 0's elsewhere.

Example

cyclic(4) is

0	1	0	0
0	0	1	0
0	0	0	1
1	0	0	0

Code

```
function A = cyclic(n)
A = diag(ones(n−1,1),1);
A(n,1) = 1;
```

drawvec

Synopsis

drawvec(v,color,s)
h = drawvec(v,color,s)

Description

drawvec(v,color,s) graphs the vector v using the color specified by the second input argument. If no second argument is specified, the default color is red. The initial point of the plot is the origin. An arrow is drawn at the terminal point of the vector. Axis is set to $[-s,s,-s,s]$. If the third input argument s is not specified, its default value is 5. The command h = drawvec(v,color,s) graphs the vector v and assigns it the graphics handle h.

Utility files

drawvec requires the ATLAST utility file tip.

Code

```
function[ h ] = drawvec(v,color,s)
if nargin == 1
    color = 'r';
end
if nargin < 3
    s = 5;
end
h = plot([0,v(1)],[0,v(2)],color)
axis([-s,s,-s,s])
axis('square')
hold on
[m,n] = size(v);
if n == 1   % Change to row vector
    v = v';
end
atip = tip(v,s);
fill(atip(1,:),atip(2,:),color)
hold off
```

editimag

Synopsis

A = editimag(B,imag,back)

Description

The command editimag(B,imag,back) edits the image represented by the matrix B that was originally created using the M-file makeimag. The user must input the matrix, the image color and the background color as arguments of **editimag**. Click on a square to toggle between the background color and the image color. Type **q** to quit.

Code

```
subplot(1,1,1)
[n,m] = size(B);
if nargin == 1
    imag = 27; back = 1;
end
if nargin == 2
    back = 1;
end
A = B;
image(A); axis('square')
while 1
    [x y z] = ginput(1);
    if abs(z) == 'q'
        break
    elseif (x >= 1) & (x < n+1) & (y >= 1) & (y < n+1)
        x = fix(x); y = fix(y);
        if A(y,x) ~= imag
            A(y,x) = imag;
        else
            A(y,x) = back;
        end
        image(A); axis('square')
    end
end
close
```

eigplot

Synopsis

    eigplot(A)

Description

eigplot(A) plots the eigenvalues of the square matrix A in the complex plane. After the plot is generated, you can zoom in on a particular eigenvalue. Move the mouse so that the pointer on the screen lines up with the eigenvalue, and then click the left mouse button. You can continue clicking the left button until you have zoomed in to a desired accuracy. You can also zoom back out by clicking on the right mouse button.

eigshow

Synopsis

eigshow(A)

Description

The command eigshow(A) command will then generate an animation showing graphically how x and Ax change as x moves around the unit circle. To start the animation, type eigshow. To move the vector x around the circle hold the left button of the mouse down and move the mouse slowly in a counterclockwise circle.

Utility files

Utility files need not be downloaded separately.

gersch

Synopsis

gersch(A)

Description

gersch(A) plots the Gerschgorin circles of the square matrix A in the complex plane. The command gersch(A,eigplot) plots the Gerschgorin circles; if eigplot equals 1 it will also plot the eigenvalues of A. One can specify a color for the plot by including as a third argument a string specifying the color. For example, the command gersch(magic(5),1,'b') will generate a plot of the eigenvalues and Gerschgorin circles of a 5×5 magic square and the plot will be drawn in blue.

gridmat

Synopsis

A = gridmat(n)

Description

gridmat(n) is the $n \times n$ matrix in which every row is [1 2 $\cdots$ n].

Example

gridmat(3) is

1	2	3
1	2	3
1	2	3

Code

```
function A = gridmat(n)
A = ones(n,1)*[1:n];
```

gschmidt

Synopsis

Q = gschmidt(A)

[Q,R] = gschmidt(A)

Q = gschmidt(A,1)

[Q,R] = gschmidt(A,1)

Description

If A is a matrix with linearly independent column vectors, then the command Q = gschmidt(A) will generate a matrix Q whose columns form an orthonormal basis for the column space of A. The matrix Q is computed using a modified version of the Gram-Schmidt process. The command Q = gschmidt(A,1) computes the matrix Q using the classical Gram-Schmidt process. This method may not be numerically stable for some matrices.

If a second output argument is included, [Q R] = gschmidt(A) or [Q R] = gschmidt(A,1), then a square upper triangular R is determined so that the A factors into a product QR.

Example

[Q,R] = [2 −1; 2 −1; 2 4; 2 4] is

```
    Q =                                      R =
          0.5000      -0.5000                      4      3
          0.5000      -0.5000                      0      5
          0.5000       0.5000
          0.5000       0.5000
```

Code

```
function [Q,R] = gschmidt(A,classic)
[m,n] = size(A);
R = zeros(n);
if rank(A) ~= n
    error('Column vectors are not linearly independent')
end
if nargin == 1 % Modified Gram-Schmidt
    Q = A;
    for k = 1:n;
        R(k,k) = norm(Q(:,k));
        Q(:,k) = Q(:,k)/R(k,k);
        R(k,k+1:n) = Q(:,k)'*Q(:,k+1:n);
        Q(:,k+1:n) = Q(:,k+1:n)-Q(:,k)*R(k,k+1:n);
    end
else % Classical Gram-Schmidt
    R(1,1) = norm(A(:,1));
    Q(:,1) = A(:,1)/R(1,1);
    for k = 2:n
        R(1:k-1,k) = Q(:,1:k-1)'*A(:,k);
        Q(:,k) = A(:,k)-Q(:,1:k-1)*R(1:k-1,k);
        R(k,k) = norm(Q(:,k));
        Q(:,k) = Q(:,k)/R(k,k);
    end
end
```

hconsec

Synopsis

A = hconsec(n)

Description

hconsec(n) is an $n \times n$ matrix whose entries in each row are consecutive integers. Furthermore the first entry in each row is the same as the number of the row. Thus the first row is [1, 2, ..., n], the second row is [2,3, ..., n+1], etc.

Example

hconsec(5) is

1	2	3	4	5
2	3	4	5	6
3	4	5	6	7
4	5	6	7	8
5	6	7	8	9

Code

```
function A = hconsec(n)
A = hankel([1:n]',[n:2*n−1]);
```

inciden

Synopsis

I = inciden(E)

Description

inciden(E) is the incidence matrix corresponding to the list of edges in the 2 × n matrix E.

Example

Suppose E is the 2 × 5 matrix

1	4	3	4	2
3	2	5	1	5

Then inciden(E) is

-1	0	0	1	0
0	1	0	0	-1
1	0	-1	0	0
0	-1	0	-1	0
0	0	1	0	1

Code

```
function I = inciden(E)
[m,n] = size(E);
if m ~= 2
    error('Requires a matrix with exactly two rows')
end
F = sort(reshape(E,1,2*n));
V = [F(1)];
for i = 2:length(F)
    if F(i) > F(i-1)
        V = [V,F(i)];
    end
end
I = zeros(length(V),n);
for i = 1:n
    I(find(V == E(1,i)),i) = -1;
    I(find(V == E(2,i)),i) = 1;
end
```

jordan0

Synopsis

A = jordan0(n)

Description

jordan0(n) is the $n \times n$ matrix of 1's on the diagonal just above the main diagonal, and 0's elsewhere.

Example

jordan0(4) is

0	1	0	0
0	0	1	0
0	0	0	1
0	0	0	0

Code

```
function A = jordan0(n)
A = diag(ones(n-1,1),1);
```

makeimag

Synopsis
A = makeimag(n,imag,back)

Description
The command makeimag(n,imag,back) lets the user create an n-by-n image, with **imag** and **back** being the image and background colors, respectively, based on the current colormap. Click with the mouse to toggle a square between background and image color. Type **q** when done.

Code

```
subplot(1,1,1)
if nargin == 1, imag = 27; back = 1; end
if nargin == 2, back = 1; end
A = ones(n)*back;
image(A); axis('square')
while 1
    [x,y,z] = ginput(1);
    if abs(z) == 'q'
        break
    elseif (x >= 1) & (x < n+1) & (y >= 1) & (y < n+1)
        x = fix(x); y = fix(y);
        if A(y,x) == back
            A(y,x) = imag;
        else
            A(y,x) = back;
        end
        image(A); axis('square')
    end
end
close
```

maxmat

Synopsis
A = maxmat(n)

Description

maxmat(n) is the $n \times n$ matrix whose (i,j) entry is the maximum of i and j.

Example

maxmat(3) is

1	2	3
2	2	3
3	3	3

Code

```
function A = maxmat(n)
G = ones(n,1)*[1:n];
A = max(G,G');
```

minmat

Synopsis

A = minmat(n)

Description

minmat(n) is the $n \times n$ matrix whose (i,j) entry is the minimum of i and j.

Example

minmat(3) is

1	1	1
1	2	2
1	2	3

Code

```
function A = minmat(n)
G = ones(n,1)*[1:n];
A = min(G,G');
```

movefig

Synopsis

A = movefig(F,M,n)
A = movefig(F,M,n,t)
A = movefig(F,M,n,t,ax)

Description

The command movefig(F,M,n) will plot the graph of a figure generated from the matrix F. The rows of points are joined by straight
lines to form the figure. The figure is then transformed by applying the matrix M to F and the transformed figure replaces
the original in the graph window. This procedure is iterated n
times. The effect is that the figure appears to be moving across
the screen. If a fourth input argument t is included, there will be
a pause of t seconds between plots. A fifth input argument of a
1×4 vector can be used to set the axis scaling. Otherwise, the
default scaling is axis([$-10,10,-10,10$]).

niceimag

Synopsis

niceimag

Description

niceimag generates a matrix for an ATLAST exercise on the singular value decomposition and digital imaging. The rank one
svd approximation to the image is displayed on the graph window. The image can then be updated by clicking the mouse. The
updates are generated by the **svdimage command** which uses the
factors from the singular value decomposition.

nulbasis

Synopsis

N = nulbasis(A)

Description

nulbasis(A) is a matrix whose columns form a basis for the null
space of A. The basis is obtained from the reduced row echelon
form of A.

Example

Suppose A is the 3×4 matrix

$$\begin{array}{rrrr} 3 & -3 & 1 & -3 \\ -2 & 2 & 1 & 2 \\ -1 & 1 & -3 & 1 \end{array}$$

Then nulbasis(A) is

```
1    1
1    0
0    0
0    1
```

Code

```
function N = nulbasis(A)
[R,jp] = rref(A);
[m,n] = size(A);
r = length(jp);
nmr = n − r;
N = zeros(n,nmr);
kp = other(jp,n);
for q = 1:nmr
    N(kp(q),q) = 1;
    N(jp,q) = −R(1:r,kp(q));
end
```

other

Synopsis

```
kp = other(jp,n)
```

Description

other(jp,n) contains the indices remaining in $1, 2, \ldots, n$ when the indices in jp are removed.

Example

```
other([ 2  3  6 ],7) is
    1    4    5    7
```

Code

```
function kp = other(jp,n)
x = 1:n;
x(jp) = zeros(1,length(jp));
kp = x(find(x));
```

plotline

Synopsis

plotline(a,b,c,s)

Description

The command plotline(a,b,c,s) plots the line ax + by = c with axis set to [−s,s,−s,s]. If the last input argument is omitted, its default value is taken to be 5.

Code

```
function plotline(a,b,c,s)
if nargin == 3
    s = 5;
end
if (a == 0 & b == 0 & c ~= 0)
    error('No such line exists')
end
if b == 0
    x = c/a;
    plot([x,x],[−s,s])
else
    y1 = (c+a*s)/b; y2 = (c−a*s)/b;
    plot([−s,s],[y1,y2])
end
axis([−s,s,−s,s])
ds = 0.2*s;
set(gca,'XTick',[−s:ds:s])
set(gca,'YTick',[−s:ds:s])
```

powplot

Synopsis

powplot(A)
powplot(A,u)

Description

The command powplot(A) is used to demonstrate geometrically the effects of applying powers of a 2×2 matrix A to any unit vector. This is done by plotting the image of the unit circle under the transformations A^k, for $k = 1, ..., 25$. If A has real eigenvalues, then an eigenvector of A is also plotted. If A has a dominant eigenvalue, then its eigenvector is the one that is plotted. If a unit vector **u** is specified as a second input argument, then the images of **u** under the powers of A are plotted.

Utility files

powplot requires the ATLAST utility file **tip**.

protate

Synopsis

protate(u,v,w,p,view,color1,color2)

Description

protate(u,v,w,p,view,color1,color2) draws the 3-tuples u, v, w as vectors with their tails at the origin. It draws **u** in color1 (with default color yellow), and it draws **v**, **w**, and the parallelogram formed by **v** and **w** in color2 (with default color red). It then rotates this picture about a line through the origin with orientation angles given by the 2-tuple **view** (with default [0,90]). The variable p is the number of seconds you want to pause between frames of the animation. All but the first three arguments are optional.

pyr

Synopsis

pyr

Description The command pyr generates the graph of an airplane on a 3-dimensional coordinate system. Three different types of rotations can be applied to the plane by entering the degrees of rotation and then clicking on the appropriate button.

Utility files

pyr requires the ATLAST utility file **pyrmover**.

randint

Synopsis

A = randint(n)
A = randint(m,n)
A = randint(m,n,k)
A = randint(m,n,k,r)

Description

randint(n) is a random n × n matrix with integer entries in the interval [−9 : 9]. randint(m,n) is a random m × n matrix with integer entries in the interval [−9:9]. randint(m,n,k) is a random m × n matrix with integer entries in the interval [−k:k]. randint(m,n,k,r) is a random m × n matrix of rank r with integer entries in the interval [−k:k].

Example

If **rand** has not yet been used during a particular MATLAB session, then the commands **randint(3,4,5)** and **randint(3,4,5,2)** produce the following matrices

-3	2	0	-5	1	1	1	2
-5	5	4	0	3	0	3	3
2	-1	-5	2	-1	0	-1	-1

Code

```
function A = randint(m,n,k,r)
test = 0; k = round(abs(k));
if nargin == 1, n = m; end
if nargin < 3, k = 9; end
if nargin < 4
      A = floor((2*k+1)*rand(m,n) − k);
elseif k == 0 & r == 0
      A = zeros(m,n);
else
      r = round(r);
      if ( r < 1 | r > m | r > n | k < 1)
            error('No such matrix exists')
      end
      p = ceil(k/r);
      while test == 0
            A = randint(m,r,p)*randint(r,n,1);
            A = A.*(abs(A) <= k);
            if rank(A) == r
                  test = 1;
            end
      end
end
```

randstoc

Synopsis

```
S = randstoc(n)
```

Description

randstoc(n) is a random $n \times n$ stochastic matrix.

Example

If rand has not yet been used during a particular MATLAB session, then randstoc(3) produces

```
0.2317    0.3401    0.3750
0.0498    0.4679    0.6000
0.7185    0.1920    0.0250
```

Code

```
function A = randstoc(n)
A = rand(n);
A = A./(ones(n,1)*sum(A));
```

rollit

Synopsis

```
rollit(n)
rollit(n,t)
```

Description

The command rollit(n) will generate a plot of a colored wheel. Press any key, and the wheel will roll across the screen for the distance n. Initially the wheel is centered at the origin. The motion is simulated by a sequence of rotations and translations. At each step the wheel is rotated in its initial position and then translated horizontally. Each successive translation moves the wheel further to the right. To slow up the motion include a second input argument t. This makes the wheel pause for t seconds every time it covers a distance of $n/60$.

rowcomb

Synopsis

```
B = rowcomb(A,i,j,c)
```

Description

rowcomb(A,i,j,c) is the result of replacing row j of matrix A by the sum of row j and c times row i.

Example

Suppose A is the 3 × 4 matrix

```
3     4    -1     2
5    -7     0    -1
2     6    -3     4
```

Then rowcomb(A,1,3,−3) is

```
 3     4    -1     2
 5    -7     0    -1
-7    -6     0    -2
```

Code

```
function B = rowcomb(A,i,j,c)
[m,n] = size(A);
if i < 1 | i > m | j < 1 | j > m
    error('Index out of range')
end
if i == j, error('Illegal row operation'), end
B = A;
B(j,:) = c*A(i,:) + A(j,:);
```

rowscale

Synopsis

B = rowscale(A,i,c)

Description

rowscale(A,i,c) is the result of multiplying row i of matrix A by the scalar c.

Example

Suppose A is the 3 × 4 matrix

```
3     4    -1     2
5    -7     0    -1
2     6    -3     4
```

Then rowscale(A,1,−2) is

```
-6    -8     2    -4
 5    -7     0    -1
 2     6    -3     4
```

Code

```
function B = rowscale(A,i,c)
[m,n] = size(A);
if i < 1 | i > m
     error('Index out of range')
end
B = A;
B(i,:) = c*A(i,:);
```

rowswap

Synopsis

B = rowswap(A,i,j)

Description

rowswap(A,i,j) is the result of swapping (i. e., interchanging) rows i and j of matrix A.

Example

Suppose A is the 3 × 4 matrix

```
3     4    -1     2
5    -7     0    -1
2     6    -3     4
```

Then rowswap(A,1,3) is

```
2     6    -3     4
5    -7     0    -1
3     4    -1     2
```

Code

```
function B = rowswap(A,i,j)
[m,n] = size(A);
if i < 1 | i > m | j < 1 | j > m
      error('Index out of range')
end
B = A;
if i == j, break, end
B(i,:) = A(j,:);
B(j,:) = A(i,:);
```

signmat

Synopsis

A = signmat(n)

Description

signmat(n) is the $n \times n$ matrix whose entries are alternately 1 and
-1, with 1 in the upper-left corner.

Example

signmat(3) is

1	-1	1
-1	1	-1
1	-1	1

Code

```
function A = signmat(n)
G = ones(n,1)*[1:n];
A = (-1).^(G+G');
```

solution

Synopsis

x = solution(A,b)

Description

x = solution(A,b) is a solution x of A x = b, and is obtained from
the reduced row echelon form of [A,b] as one does by hand.

Example

Suppose A is the 3 × 4 matrix

```
    0    -5     4     2
    0     3    -4     4
    5     3    -5     1
```

and b is the 3 × 1 matrix

```
    4
   -3
   -1
```

Then solution(A,b) is

```
    0.4750
   -0.5000
    0.3750
         0
```

Code

```
function x = solution(A,b)
[R,jp] = rref([A,b]);
[m,n] = size(A);
r = length(jp);
if jp(r) == n+1
    x = [];
else
    x = zeros(n,1);
    x(jp) = R(1:r,n+1);
end
```

svdimage

Synopsis

svdimage(A,U,S,V,k,cmap,flag)

Description

The command **svdimage(A,U,S,V)** will display the image represented by the matrix A based on the current colormap. It will
then step through the svd approximations to the image (starting
with the rank 1 approximation) each time the left mouse button
is clicked in the figure window.

The parameters U,S,V are the factors of the singular value decomposition of A.

The command svdimage(A,U,S,V,k) will start with the rank k approximation of the image.

To change from the default colormap, include a new colormap matrix as a sixth input argument.

To suppress the display of the original image until the end, include a seventh input argument which may have any value.

svdimage(A) is the same as image(A). It will display the original image.

svdshow (This file is being replaced by an svd option for eigshow.)

Synopsis

svdshow(A)

Description

The command svdshow(A) command will generate an animation showing graphically how Ax and Ay change as a pair of orthogonal vectors x and y are rotated around the unit circle. The angle between Ax and Ay is also displayed.

To start the animation, type svdshow(A). To move the vectors x and y around the circle, hold the left button of the mouse down and move the mouse slowly in a counterclockwise circle.

When Ax and Ay are orthogonal, then x and y are right singular vectors of A and Ax and Ay are the left singular vectors multiplied by the singular values.

Utility files

svdshow requires the ATLAST utility file svdshowf.

transfor

Synopsis

transfor

Description

The command transfor invokes the ATLAST Transformer. This utility allows one to see graphically the effects of linear transformations on geometric figures.

An initial image is selected from a pull-down image menu. One can then select a transformation, either Rotation, Reflection, or Diagonal. The user must input either the degrees of rotation or reflection or the scaling factors of the diagonal transformation in the appropriate box in the graphics window. Press return or click the appropriate transformation button to perform the transformation. The image of the figure under the selected transformation will be displayed in the subplot labelled "Current Image".

Separate subplots display the initial version of the figure, the current transformed version of the figure, and the previous transformed version.

A Freeze button will copy the current image to a fourth target subplot. The Undo button will undo the effects of the last transformation that was applied. The Restart button resets the current and previous image subplots to the initial image. The Current Transformation button will display the matrix of the current composite transformation in the graphics window.

The Mystery pull-down menu is used to generate mystery transformations. The image of the initial figure under the mystery transformation is displayed in the target subplot. Mystery transformations are composites of rotations, reflections, and diagonal scalings. Angles of rotation or reflection are multiples of either 45 or 60 degrees. The diagonal scaling factors are either 1, 2, or 1/2. The level of the mystery transformation is the same as the number of composite transformations used in its construction. Level 3 mystery transformations are often difficult to guess. To help, a hint is displayed in the MATLAB command window.

Utility files

transfor requires the ATLAST utility files **imager**, **mystery**, **options**, **PlotImag**, **trans**, **stretch**.

walk

Synopsis

walk(steps)

walk(steps,t)

Description

The command **walk(steps)** will generate a plot of a stick figure. Press any key and a stick figure will walk across the screen. The input argument, **steps**, determines how far it walks. The movement of the upper body is a successions of horizontal translations. The leg movements are rotations about the origin followed by horizontal translations so that the legs keep up with the body. (A desirable property for walking.) A second input argument t will cause the figure to pause t seconds after each incremental movement. If no second input argument is included, the default pause is 0. To avoid pauses altogether, set the second input argument equal to 'off', e.g., **walk(10,'off')**.

References

1. Cowen, Carl C., "A Project on Circles in Space." In *Resources for Teaching Linear Algebra*. Edited by Carlson, David, Charles R. Johnson, David Lay, A. Duane Porter, Ann Watkins, and William Watkins. Washington, D.C.: MAA, 1996.

2. Hanselman, Duane, Bruce Littlefield, and the staff at the MathWorks, Inc., *The Student Edition of MATLAB, Version 4*. Upper Saddle River, New Jersey: Prentice-Hall, 1995.

3. Leon, Steven J., *Linear Algebra with Applications, Fourth Ed.*, Upper Saddle River, New Jersey: Prentice-Hall, 1994.

The "Circles in Space" project in Chapter 8 is based on reference [1]. Our Appendix A is taken directly from the MATLAB Appendix in reference [3]. This Appendix covers the basics of using MATLAB. Reference [2] provides a more complete guide to MATLAB.